An Illustrated Chinese-English Guide for Biomedical Scientists

An Illustrated Chinese-English Guide for Biomedical Scientists

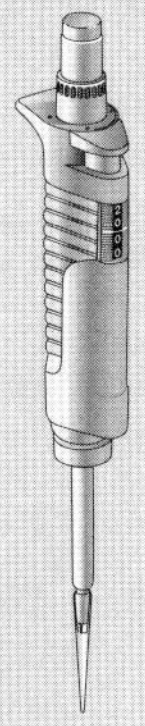

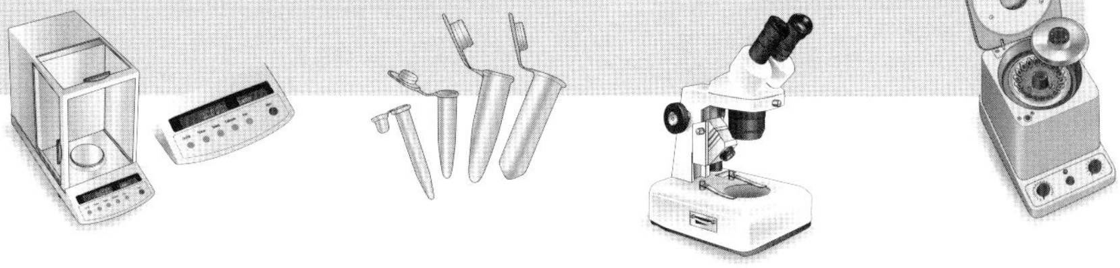

James M. Samet

Weidong Wu

Yuh-Chin T. Huang

Xinchao Wang

COLD SPRING HARBOR LABORATORY PRESS

Cold Spring Harbor, New York

An Illustrated Chinese-English Guide for Biomedical Scientists

Printed in China

Acquisition Editor and Publisher:	John Inglis
Developmental Editor:	David Crotty
Development Manager:	Jan Argentine
Project Coordinator:	Joan Ebert
Production Editor:	Mala Mazzullo
Desktop Editor:	Danny deBruin
Book Designer and Production Manager:	Denise Weiss
Cover Designer:	Ed Atkeson/Denise Weiss
Front cover artwork and illustrations:	Hans Neuhart/EIG

Front Cover: Illustrations of an adjustable micropipetter, an analytical balance, microcentrifuge tubes, a stereo dissecting microscope, and a microcentrifuge.

Library of Congress Cataloging-in-Publication Data

An illustrated Chinese-English guide for biomedical scientists /
James M. Samet ... [et al.].
p. cm.
In Chinese and English.
Includes index.
ISBN 0-87969-701-6 (cloth : alk. paper) -- ISBN 0-87969-648-6
(pbk. : alk. paper)
1. Molecular biology--Research--Terminology. 2. Biological
laboratories--Terminology. 3. Medical laboratories--Terminology. 4.
Biological laboratories--Equipment and supplies. 5. Medical
laboratories--Equipment and supplies. I. Samet, James M.
QH506.I46 2003
572.8'072--dc22

2003070118

10 9 8 7 6 5 4 3 2 1

All Cold Spring Harbor Laboratory Press publications may be ordered directly from Cold Spring Harbor Laboratory Press, 500 Sunnyside Blvd., Woodbury, N.Y. 11797-2924. Phone: 1-800-843-4388 in Continental U.S. and Canada. All other locations: (516) 422-4100. FAX: (516) 422-4097. E-mail: cshpress@cshl.edu. For a complete catalog of all Cold Spring Harbor Laboratory Press publications, visit our World Wide Web Site http://www.cshlpress.com.

For Jingyi and Richard;

for Xiaobo and Boyu;

for Mingyi, Jennifer and Megan;

and for Ryan and Madeleine:

May they live in a world where language is

not a barrier to the exchange of ideas

Contents

FOREWORD
Simplified Chinese, ix
Complex Chinese, xi
English, xiii
INTRODUCTION
Simplified Chinese, 1
Complex Chinese, 5
English, 9
U.S. TRAVEL INFORMATION FOR CHINESE SCIENTISTS
Simplified Chinese, 13
English, 19
ILLUSTRATED LABORATORY TERMINOLOGY
Measuring, 25
Processing, 31
Culturing, 39
Documenting, 43
Detecting & Assaying, 46
Separating & Fractionating, 50
Supplies & Equipment, 53
Safety, 57
COMPREHENSIVE LABORATORY TERMINOLOGY, 63
ILLUSTRATIONS INDEX, 113

序言

利器，好书

古人云：工欲善其事，必先利其器。本书介绍的是分子生物学技术这一工具，且它本身就是掌握这些技术的有力工具。

我非常感谢冷泉港实验室出版社出版这样一本好书。因为该书不仅是一般意义上的精确高效的工具书，而且是第一本主要由中国作者专为中国科学工作者所撰写的中英文对照的工具书。

本书的对象是刚刚涉足分子生物学这一新的领域的生命科学家。对生命的认知与对生物的利用，是人类历史的重要组成部分。然而，从农业的曙光初现伊始，包括中华民族祖先在内的人类先祖在生物学领域硕果累累，我们对生命的奥秘却仍不甚了解。对生命的探索，在重新发现遗传规律的二十世纪才真正开始。

我向所有致志于生命科学的同道们祝贺！探索生命奥秘，其乐无穷！1953年，James Watson 与Francis Crick 共同发现的DNA双螺旋结构，给生命科学带来了突破性进展。我们这一代非常荣幸地见证了这些科学巨变。由此重大发现派生的生物技术，已经对人类社会和生活产生了深刻的影响。我与我的同事们都为能生活在这一历史时期深感欣喜。我们孜孜以求，期望在科学的征程上能有更多的收获。我们的抱负是充分发挥自己的潜能，做到事业有成。这些目标的实现，离不开好的研究工具。这本书就是要把这些研究工具介绍给世界各地的华人读者。

尽管我们在专业知识方面做好了进入科研新时代的准备，可我们还面临在交流及理解方面的障碍。在应用第二语言上我们也许会感吃力。作为该书作者的实验指导人员及同事们正是理解体谅了我们这方面的难处。为此，特向他们深表谢枕。该书编写具有很强的针对性。我极力推荐该书的原因就是因为它的科学性，可读性及新颖的表述。

该书可作为学生的英语教科书。对那些在中国国内掌握了实验技术，却难于恰如其分地在英语讲座或会谈中表达自己的学者来说，这本书可谓不可多得。此书也可作为那些悉心培训与指导华人科学家的人员的教材。这些人员在培训母语非英语人员时，注重声情并茂及中肯贴切的表达。

我确信国内读者会发现这是一本很有价值的书。我特将此书推荐给那些期望去国外学习和培训的同胞。倘若你正在和将要从事某方面的生物科学实验室工作，这本手册将会成为你的知心朋友，令你爱不释手。我相信，在不久的将来，这本手册将同其它诸如《分子克隆》等冷泉港实验室出版的巨著一道，出现在华人学生和学者工作的实验室里。值此，我祝贺读者有幸获得这本书。

中国科学院北京基因组研究所
杨焕明　　所长

序言

利器，好書

古人云：工欲善其事，必先利其器。本書介紹的是分子生物學技術這一工具，且它本身就是掌握這些技術的有力工具。

我非常感謝冷泉港實驗室出版社出版這樣一本好書。因為該書不僅是一般意義上的精確高效的工具書，而且是第一本主要由中國作者專為中國科學工作者所撰寫的中英文對照的工具書。

本書的對象是剛剛涉足分子生物學這一新的領域的生命科學家。對生命的認知與對生物的利用，是人類歷史的重要組成部分。然而，從農業的曙光初現伊始，包括中華民族祖先在內的人類先祖在生物學領域碩果纍纍，我們對生命的奧秘卻仍不甚了解。對生命的探索，在重新發現遺傳規律的二十世紀才真正開始。

我向所有致志于生命科學的同道們祝賀！探索生命奧秘，其樂無窮！1953年，James Watson 与Francis Crick 共同發現的DNA雙螺旋結构，給生命科學帶來了突破性進展。我們這一代非常榮幸地見證了這些科學巨變。由此重大發現派生的生物技術，已經對人類社會和生活產生了深刻的影響。我与我的同事們都為能生活在這一歷史時期深感欣喜。我們孜孜以求，期望在科學的征程上能有更多的收獲。我們的抱負是充分發揮自己的潛能，做到事業有成。這些目標的實現，离不開好的研究工具。這本書就是要把這些研究工具介紹給世界各地的華人讀者。

盡管我們在專業知識方面做好了進入科研新時代的准備，可我們還面臨在交流及理解方面的障礙。在應用第二語言上我們也許會感吃力。作為該書作者的實驗指導人員及同事們正是理解體諒了我們這方面的難處。為此，特向他們深表謝枕。該書編寫具有很強的針對性。我極力推荐該書的原因就是因為它的科學性，可讀性及新穎的表述。

該書可作為學生的英語教科書。對那些在中國國內掌握了實驗技術，卻難于恰如其分地在英語講座或會談中表達自己的學者來說，這本書可謂不可多得。此書也可作為那些悉心培訓與指導華人科學家的人員的教材。這些人員在培訓母語非英語人員時，注重聲情并茂及中肯貼切的表達。

我確信國內讀者會發現這是一本很有价值的書。我特將此書推荐給那些期望去國外學習和培訓的同胞。倘若你正在和將要從事某方面的生物科學實驗室工作，這本手冊將會成為你的知心朋友，令你愛不釋手。我相信，在不久的將來，這本手冊將同其它諸如《分子克隆》等冷泉港實驗室出版的巨著一道，出現在華人學生和學者工作的實驗室里。值此，我祝賀讀者有幸獲得這本書。

中國科學院北京基因組研究所

楊煥明　所長

Foreword

A powerful tool, a wonderful book

As an ancient Chinese proverb goes: No good job can be done without a good tool. This book is about a powerful tool—the technology of molecular biology—and is itself a wonderful tool for learning this technology.

I am very grateful to Cold Spring Harbor Laboratory Press for publishing it. Not only because it is a valuable book in a general sense, but also because it is written in Chinese for Chinese scientists by a largely Chinese author team.

The book is intended to help those who are entering the new world of the life sciences. The process of understanding life and the utilization of other living organisms has been an integral part of the history of man. However, even though there have been great achievements in biology, including those of our Chinese ancestors, since the dawn of agriculture and the domestication of animals, the exploration of the secrets of life did not substantially begin until the first weeks of the last century when the laws of inheritance were rediscovered.

I congratulate all new recruits to the life sciences on their choice of career. You will find that investigation of the secrets of life is wonderfully satisfying. I am blessed to have witnessed, with other scientists of my generation, the remarkable breakthroughs that have been made since the discovery of the DNA double helix by James Watson and Francis Crick in 1953. I also share with my colleagues the satisfaction and joy of living in a period of history in which biotechnology, based on the double helix, has made such a great impact on our society and our lives. We push forward, expecting to harvest still more from science and with the ambition of realizing our own potential and making successful careers. We need good tools to achieve all that. This book is about those tools, and it brings them to Chinese readers everywhere.

Even though we are scientifically well prepared to enter this new era of research, we still face difficulties in communication and learning. We may lack confidence in our abilities in a second language and therefore much appreciate supervisors and colleagues who understand our difficulties. This book is written for us with the best of intentions. I strongly recommend it for the excellence of its science and its readable and creative language.

It is a student's textbook in English, invaluable for those who may have learned a technique in China but have difficulties in finding the right words and the right way to express themselves when they prepare for an interview or a seminar in English. But it will also become a teacher's textbook, for considerate supervisors of Chinese trainees who believe in the kind and empathetic art of choosing the right English words and expressions when training somebody who is not a native speaker.

I'm also sure that readers inside China will find this book extremely valuable. I especially recommend it to those who wish to go abroad for further studies or training. If you are working or preparing to work in a laboratory in any field of the life sciences, you will find this handbook to be like a friend with whom you always want to spend time. I am confident that it will soon be found alongside that other famous book from Cold Spring Harbor, *Molecular Cloning,* in those numerous laboratories where Chinese students and scientists are performing their experiments. I congratulate you, the reader, on your good fortune in acquiring it.

Huanming Yang
Director
Beijing Genomics Institute, China

前言

编写本指南的想法产生于作者所在的实验室。这里和美国的其它实验室一样，中国科学工作者与他们的美国同事一起工作。本书作者所从事的生物医学方面的研究工作是一个建立于个人观点交流及观点顺畅沟通基础上的活跃领域。所以，语言交流上的障碍会严重影响研究的进展。

我们发现，在美国实验室工作的中国访问学者，通常都有很好的英语功底；然而，从事科研的实验室语言环境与TOEFL 和其它英语考试所测试的又大不相同，因为实验研究已经变得如此专业化，以至于相邻实验室的工作人员使用完全不同工作词汇也并非罕见。如果连一个普通的美国科研人员偶尔也感到需要加脚注才能理解走廊对面实验室的对话，那么语言障碍只能加重这个问题。换句话来讲，实验室内交流意见和看法对整个科研过程至关重要，但即使没有语言障碍，挑战仍然存在。作为科研工作者，我们常常不得不中断实验室的讨论而去澄清一些常见术语的意义，因而分散了讨论的主题。当然，我们可以利用一些高质量的英汉技术词典。但在实验室里，除非遇到极不常用的词汇我们才会求助于词典。查词典毕竟是件烦琐的事情。那么我们所需要的应该是一个具有实用价值的实验室专用的翻译工具。

这种情景令人联想起当时启发已故自然学家，现代观鸟之父，罗杰托瑞皮特森，创作他的《北美东部、中部野外观鸟指南》(*Field Guide to the Birds of Eastern and Central North America)* 的情形。在皮特森的野外观鸟指南一书出现以前，众多鸟类学参考书冗长而不实用。皮特森的革新性创造就在于它的绝对实用性。我们当中不少人曾是业余观鸟者，我们深知皮特森的创举。不久前，我们注意到，实验室里的许多语言僵局实际上完全可以通过在试验台上画一张最简单的草图而解决。于是我们着手编写一本高效实用的指南，以备实验桌边的随手参考以及英汉实验室术语翻译之用。

正如皮特森的鸟类图例所示及实验台示意图所强调的一样，当交流既强调速度又强调全面的时候，什么也比不过图片。 在设计此指南时，为便于查找，我们设法把图解作为本书的中心内容并按主题进行分类以发挥图解的最大作用。本书尽量做到对一般实验人员直观明了，比如把烧瓶和倾口烧杯一起描述，而把工作服和安全眼镜归为同一主题。当然，实验室使用的术语并非每个都可用制图轻易描述，所以本书也提供了一个词汇表。而且，与图解主题关联的术语也在本书最后的主题部分出现。

经咨询科研人员和图书管理人员，我们认为本指南须有语音部分。大家知道，很多科研信息是通过讲座以及座谈的方式来传递的，所以对于演讲者而言，正确掌握所讨论术语的发音至关重要。为此，我们为本指南中的每一术语制作了网站语音文件 (http://chinese-english.cshl.org)。

在国外访问常常会面临远远超出语言问题的各种挑战。本指南的每一位中国作者都有到美国实验室的旅行经历，因此，他们对有关旅行准备、到达美国、房屋租赁以及各种文件手续等非常了解，他们愿将所累积的经验与准备到美国实验室工作的中国科学工作者们一道分享。真切希望本指南中包括的这些实用性建议能够有益于未来的访问者。

本书起源于作者面对那些高雅的鸟类插图以及实验台边的随手草图所产生的灵感。我们感谢将此灵感变成现实的人们。编写第一版《分子生物学实验室工作汉英图解指南》，很大程度上是凭一种直觉，好在冷泉港出版社的编辑人员给予了审慎的指点。毫无疑问，本书很多方面都有待进一步改进，因此我们非常感谢读者的任何建议，以期在再版时涵括您的意见。我们的目的很简单，那就是促进实验室里中美科学工作者之间的交流。只有读者您才能评价我们所完成的内容。

前言

編寫本指南的想法產生于作者所在的實驗室。這里和美國的其它實驗室一樣，中國科學工作者与他們的美國同事一起工作。本書作者所從事的生物醫學方面的研究工作是一個建立于個人觀點交流及觀點順暢溝通基礎上的活躍領域。所以，語言交流上的障礙會嚴重影響研究的進展。

我們發現，在美國實驗室工作的中國訪問學者，通常都有很好的英語功底；然而，從事科研的實驗室語言環境与 TOEFL 和其它英語考試所測試的又大不相同，因爲實驗研究已經變得如此專業化，以至于相鄰實驗室的工作人員使用完全不同工作詞匯也并非罕見。如果連一個普通的美國科研人員偶爾也感到需要加腳注才能理解走廊對面實驗室的對話，那么語言障礙只能加重這個問題。換句話來講，實驗室內交流意見和看法對整個科研過程至關重要，但即使沒有語言障礙，挑戰仍然存在。作爲科研工作者，我們常常不得不中斷實驗室的討論而去澄清一些常見術語的意義，因而分散了討論的主題。當然，我們可以利用一些高質量的英漢技術詞典。但在實驗室里，除非遇到極不常用的詞匯我們才會求助于詞典。查詞典畢竟是件煩瑣的事情。那么我們所需要的應該是一個具有實用价值的實驗室專用的翻譯工具。

這种情景令人聯想起當時啓發已故自然學家，現代觀鳥之父，羅杰托瑞皮特森，創作他的《北美東部、中部野外觀鳥指南》（*Field Guide to the Birds of Eastern and Central North America*）的情形。在皮特森的野外觀鳥指南一書出現以前，眾多鳥類學參考書冗長而不實用。皮特森的革新性創造就在于它的絕對實用性。我們當中不少人曾是業余觀鳥者，我們深知皮特森的創舉。不久前，我們注意到，實驗室里的許多語言僵局實際上完全可以通過在試驗台上畫一張最簡單的草圖而解決。于是我們著手編寫一本高效實用的指南，以備實驗桌邊的隨手參考以及英漢實驗室術語翻譯之用。

正如皮特森的鳥類圖例所示及實驗台示意圖所強調的一樣，當交流既強調速度又強調全面的時候，什么也比不過圖片。 在設計此指南時，爲便于查找，我們設法把圖解作爲本書的中心內容并按主題進行分類以發揮圖解的最大作用。本書盡量做到對一般實驗人員直觀明了，比如把燒瓶和傾口燒杯一起描述，而把工作服和安全眼鏡歸爲同一主題。當然，實驗室使用的術語并非每個都可用制圖輕易描述，所以本書也提供了一個詞匯表。而且，与圖解主題關聯的術語也在本書最后的主題部分出現。

經咨詢科研人員和圖書管理人員，我們認爲本指南須有語音部分。大家知道，很多科研信息是通過講座以及座談的方式來傳遞的，所以對于演講者而言，正确掌握所討論術語的發音至關重要。爲此，我們爲本指南中的每一術語制作了网站語音文件（http://chinese-english.cshl.org）。

在國外訪問常常會面臨遠遠超出語言問題的各种挑戰。本指南的每一位中國作者都有到美國實驗室的旅行經歷，因此，他們對有關旅行准備、到達美國、房屋租賃以及各种文件手續等非常了解，他們愿將所累積的經驗与准備到美國實驗室工作的中國科學工作者們一道分享。真切希望本指南中包括的這些實用性建議能夠有益于未來的訪問者。

本書起源于作者面對那些高雅的鳥類插圖以及實驗台邊的隨手草圖所產

生的靈感。我們感謝將此靈感變成現實的人們。編寫第一版《分子生物學實驗室工作漢英圖解指南》，很大程度上是憑一种直覺，好在冷泉港出版社的編輯人員給予了審慎的指點。毫無疑問，本書很多方面都有待進一步改進，因此我們非常感謝讀者的任何建議，以期在再版時涵括您的意見。我們的目的很簡單，那就是促進實驗室里中美科學工作者之間的交流。只有讀者您才能評价我們所完成的內容。

Introduction

THE CONCEPT FOR THIS GUIDE WAS DEVELOPED in the authors' laboratory where, as in other research labs across the United States, Chinese scientists work side-by-side with their American colleagues. The authors' business is biomedical research, an endeavor that thrives on continuous communication and the fluid exchange of ideas. Therefore, communication problems are, effectively, an impediment to research.

In our experience, the typical Chinese visitor to an American lab has an excellent command of the English language. However, research labs operate in a linguistic domain that is quite different from that tested by TOEFL and other language proficiency exams. Because research has become so specialized, it is not uncommon for the staff in laboratories adjacent to each other to have substantially different working vocabularies. If the average American researcher has on occasion felt the need for subtitles to follow the dialog in the lab across the hall, linguistic barriers only magnify the problem. Said another way, communicating concepts and ideas in the lab is vital to the process, but challenging even when no language barrier exists. As scientists, we have found interruptions in laboratory discussions caused by confusion over the meaning of common terms to be recurring and tiresome distractions. Very fine technical English-Chinese dictionaries exist, of course, and they are invaluable tools. However, we found that in practice, dictionaries are too cumbersome to use for all but the most intractable translation difficulties in the lab. What was needed, then, is a lab-specific translation tool with practical appeal.

In some ways, this situation was reminiscent of the circumstances which inspired the late naturalist Roger Tory Peterson, father of modern birdwatching, to create his *Field Guide to the Birds of Eastern and Central North America*. Prior to Peterson's invention of the field guide, ornithologic references were unwieldy tomes that were too difficult for practical use. The genius of Peterson's revolutionary invention is in its sheer practicality. Some of us being amateur birdwatchers, Peterson's example was not lost on us, and it was not too long before we took notice of the fact that many a linguistic impasse in the lab could be overcome with a primitive illustration hastily drawn on the bench paper. We thus took on the task of developing an efficient and practical guide that can be kept at the ready on the bench for easy reference and translation of laboratory terms between Chinese and English.

As Peterson's bird illustrations had shown and the bench paper cartoons had reinforced, when it comes to communicating with speed and universality, nothing beats a picture. And in designing this guide, we have tried to exploit the power of illustrations to maximum effect by making them the centerpiece of the guide and organizing them into themes for easier lookup. The themes were designed to be intuitive to the average laboratory worker, so that flasks can be found near the beakers and the lab coat is close to the safety glasses. Of course, not every term used in the lab is easily depicted with a drawing and, therefore, a glossary of terms has also been included. But here again, glossary terms that are commonly associated with an illustrated theme also appear at the end of the theme sections.

Consultations with scientists and librarians made us aware of the need for an audio component to this guide. Much scientific information is transmitted through giving lectures and having conversations, so it is obviously important for a speaker to know the correct pronunciation of the terms that are being discussed. To accomplish this, we have prepared a website (http://chinese-english.cshl.org) featuring audio files of every term found in this guide.

Visiting any foreign country poses challenges that extend well beyond language issues. Each of the Chinese-speaking authors of this guide has been through the experience of traveling to the U.S. to work in a laboratory. Consequently, the issues surrounding travel preparations, arrival, housing, documentation, etc., are quite familiar to them and they are in a position to share the cumulative knowledge of their experiences with other Chinese scientists planning work in an American lab. Their practical advice, included in this guide, is offered in the sincere hope that it will be useful to these prospective visitors.

We express our appreciation to everyone who made this project, inspired by glorious bird illustrations and humble drawings on bench paper, a reality. In writing this first edition of *An Illustrated Chinese-English Guide for Biomedical Scientists*, we have largely followed our instincts, which were, thankfully, tempered by the sage counsel of our handlers at Cold Spring Harbor Laboratory Press. Undoubtedly, there are improvements to be made in every aspect of this work and we will be grateful for any and all suggestions from the users of this guide for inclusion in future editions. Our goal has been simple: to facilitate communications in the laboratory between Chinese and American scientists. Only you, the reader, can judge the extent to which we have succeeded.

X.W.
W.W.
T. Y. H.
J.M.S.

中国学者旅美须知

在美国生活多年，亲身经历了开始异乡生活时的不便。为了使后来者少走弯路，尽快适应新的环境，全身心投入紧张的工作，我们将根据自己初到美国时的体验，在本书正文之前就几个常见的问题粗略提出自己的意见，仅供读者参考。

到美国前，要做哪些准备工作？在美国使(领)馆获得允许入美的签证后，接下来就是做些入美前的准备工作。准备工作因人而异，但不外乎从生活，工作及未来可能涉及到的身份转换等几个方面进行准备。

旅行前的准备工作：

1. 要尽可能在赴美前落实住处。联系住处有多种渠道：可以请已在美国的朋友，同学，或老板与附近的租房公司或个人联系；也可请学校的留学生中心或中国学生/学者联谊会给予帮助。另外，你不妨尝试在因特网上搜索你感兴趣的住房。例如，根据你美国工作地所在的州及城镇，你可以在一些公众网站上(如http://www.apartments.com)，通过限定住房与工作单位的距离或房租的高低，搜寻你感兴趣的公寓。然后通过电子邮件或传真与有关租房办公室预约，联系。

2. 联系接机人。刚到美国，人生地疏，最好有人去机场接你。你的同学，朋友或老板也许会主动把你接回来。另外，在大多数美国高校里，有不同形式的中国学生/学者联谊会，他们将迎接新生作为年度的一项重要事情。通过浏览所在学校的官方网站，使用有关搜索工具，你也许很容易就能联系到中国学生会。告诉他们飞机的航班号及预定到达时间。你若携带多件行李，最好在临行前换些零用美元，作为到美国后中转飞机时租借行李车等的费用。有关机票费用及航班飞行情况，可在一些相关网站上了解，比较。如：http://www.cheapticket.com；http://www.priceline.com。

3. 去美国时，要带些什么生活及工作用品？这是许多人来美前常问的一个问题。若条件允许，先浏览一下 http://www.yahoo.com，http://www.travel.com，或http://www.cities.com 等公众网站，输入你要去的州或城市名，你就可了解该地区的大致情况，包括天气及人情地理等。

在你将入住的公寓里，会配有最基本的生活设施，如冰箱，炉灶，热水器等。随行李携带1-2件烹饪用具，可能会为你应付最初的异乡生活有帮助。当然，你也可在到美国后，去附近的百货店里购买。如果你能带些简单的卧具，或许能帮你摆脱初来乍到时遇到的窘迫。要知道多数公寓不备家具，需你日后逐渐添置。别忘了随身携带些药品，这因个人健康情况而定，以应付急性消化道，呼吸道疾病为主。要有孩子随行时，应带些有效的儿童药物。如果孩子为适学儿童，别忘带上由检疫机关颁发的预防接种证书。这是孩子在美国入学所必需的。说到工作用具，你可考虑带上 1-2本权威专业书或工具书，作为适应最初工作时的参考。你不必花费过多的精力和财力为自己未来的上司或同事购买礼品。贵重的礼品往往让接受方感到不自在。小巧且有民族特色的工艺品，较为对方所欣赏。如果你有在美国多呆一段时间的打算，那么临行前最好办理一些重要文件的公证，如无犯罪记录公证，学位公证，学分成绩公证，出生公证，结婚公证，亲属关系公证，及奖励证书公证等，以便将来自己转换身份或邀请家人来美探亲时用。

在飞行途中，你将会遇到许多已在在美国生活多年的华人，可借此机会向他们粗浅了解你所在州或学校的大致情况及一些生活注意事项，风俗习惯。请记住，你在飞机上要做的一项重要事情是填写I-94卡。I-94 卡虽小，可其作用却非常重要！此卡也称“出入境卡”，是非移民签证人员入美的登记卡。在你通过美国海关时，有关官员会在卡上加戳，注明你入关时的身份，入港所在地等。当你去第三国旅行时，更要随身携带，以备返美时在海关验证时用。

到达美国后优先做的工作：

万事开头难。初到美国，你将面对的是一个陌生而新奇的生活环境。语言交流吃力，不熟悉对方的风俗习惯及工作环境卫生安全规程，及可能的交通困难，会给你带来一些生活上的不便。此时要有耐心及毅力，勇敢迎接新的挑战。在到美后的一段时间里，有一些事情需你尽快办理。

1. 报到，登记。到了目的地后，你要尽快去工作单位报到。在那里，秘书会把你的一些重要文件的复印件存档，如护照，签证，I-94卡，I-20表或IAP-66表等。你也许会从那里拿到一些重要的表格，如工资发放，免税，及健康保险等。你要尽快认真填写上交。此外，你需到你所在学校留学生中心报到。同样也要上报一些上述重要文件的复印件，同时你需要填写一些个人背景资料。在那里，你会得到一些有用的生活信息资料及学校环境，概况介绍。

除在单位报到外，来自大陆的学生学者，还需到中国住美国使馆报到。当然最便捷的方式是在线报到。该网址是http://www.china-embassy.org。同时你还会在该网页上了解一些中国使馆的情况及留学生政策。

2. 申请社会安全号。你可能会在上述的留学生中心获得申请社会安全卡的信息及表格。有些学校会即时发给你一个临时社会安全号。社会安全卡虽不能作为你的身份证件，但在很多场合你需出具，如安装电话，银行开户，学习驾车等。若需你自己申办社会安全号，可浏览http://www.ssa.gov/online/，了解有关信息，包括离你居住地最近的社会安全号申请部门的地址。

3. 了解课题研究计划及工作有关规章制度。到美后，要尽快了解老板给你的课题计划，熟悉有关技术，操作规程及日常实验室管理运作程序。同时，你也可向老板谈及你生活上的某些现存的临时性困难，寻求可能的帮助。

4. 健康保险。加入健康保险计划，可减轻你看病，住院及其他医药费用所造成的经济负担，同时也是满足美国政府对学生，学者加入健康保险的要求。多数情况下，雇人单位会为你支付用于健康保险的费用。个别情况下，则由你本人支付这部分费用。你可去所在校的留学生中心咨询有关信息。另外，你也可参考http://www.student-resources.net里的有关内容，选择一种较好的健康保险服务计划。

5. 建立银行帐户。出於安全，方便的考虑，到美国后你要尽快到附近的银行申请建立自己的帐户。届时，银行职员会送给你一些附有你账号的支票。你可将支票的复印件，连同有关表格一并送交工资管理部门。之后，你的工资将会定期自动存入到你新的帐户内。你可通过该银行自己专门的服务网站，直接即时了解你帐户收支情况。

6. 安装电话。电话是你与家人，朋友及同事交流联系的一种重要的工具。不同的地区，会有不同的电话公司负责服务。携带有关证件及押金，去附近的电话公司办事处申请电话号码。关于选用长途电话服务的问题，我们认为买电话卡可能更经济方便。目前，有很多电话卡网上服务。你可直接在网上订购电话卡。

7. 办理探亲手续。若申请家庭成员来美探亲，建议先去留学生中心了解有关手续，再向周围有类似经历的人打听。你将邮寄给亲属的文件大致有以下几类： 你在美工作或生活的证明文件(复印件)，如IAP-66，I-20， I-94等。所在工作单位出具的雇佣证明，包括工资，职位及工作期间等。存款或财产证明文件，包括银行存款证明及最近的工资支付条。给探亲成员的邀请信，要中英文各一份。最好使用有单位名称的信纸。其他材料包括亲属关系证明，包括有关公证文件或最新的合影照片。

8. 学习驾车。待住处安置妥当，工作有了头绪后，你的下一个目标也许是买车，掌握驾驶技术吧？这既能方便你的工作，又可丰富你的个人生活。首先，你要获得临时性驾驶学习执照。要拿到该执照，你需通过所在州的有关交通法规及交通信号识别的理论考试。有关资料(驾驶员手册)可去你附近的车辆管理部门索取。另外，你也可在网页上浏览驾驶员手册，该网址为http://www.dmv.org。拿到驾驶学习执照后，你就要准备买车及学车了。买车前要向周围的人了解几种主要车型的大致性能。有了意向后，再去着手发现你感兴趣的车。买车信息可在地方性报纸，单位墙报，社区网站，及一些特殊的车辆市场网站上获取，如 http://www.usedcars.com 或 http://www.kbb.com 等。在这些网站上，输入你感兴趣车辆的有关信息，你还可事先了解该车的大致价格。对于旧车，最好在买前让修车行做一下机械检查，以发现主要问题。买车后，你可请人教你开车，也可去专门的汽车驾校学习驾驶。最后在汽车管理部门，通过道路测试，拿到正式驾驶执照。

9. J–1签证的豁免。持有J-1 签证的交流学者，在美国停留三年后，须返回所在国服务。由于特殊原因需在美国滞留超过签证规定时间，持J-1者须向美国移民局申请豁免返回原所在国居住两年的限制。申请豁免手续的 办理，可参考有关中国使馆及美国国务院的有关规定程序进行。这些相关网站是：http://www.sino-education.org ; http://www.travel.state.gov 。

以上仅是根据我们自己的体验，就刚到美国时几个常遇到的问题，谈一下自己的看法。难免有不当及局限之处。如果这些能对您到美后适应新生活有点滴帮助，那将令我们深感欣慰。

U.S. Travel Information for Chinese Scientists

Once your visa is issued by the United States embassy or consulate, there are still many things that you will need to prepare. Keep in mind that there is always an initial period of inconvenience when you live in a foreign country. Based on our own experiences, here we offer some advice to minimize the initial difficulties that may be encountered by those who are planning to study or work in America.

BEFORE TRAVELING

1. Finding a place to live. Try to consult a friend or mentor in America in your search for living space before you book your air ticket. If you are heading to a university, contact their housing office for information on any housing options that they offer. Most universities also have an international center or office that provides foreign students with information. There may also be a Chinese student/scholar organization on campus or in the community. The following Web site may be helpful in arranging a place to live in the U.S.: http://www.apartments.com. This site allows searches according to location, price range, and other features.
2. Travel and ground transportation upon arrival. Several well-known Web sites are available for searching flight information and purchasing tickets (http://www.cheaptickets.com, http://www.travelocity.com, etc.). It is best to contact someone at your destination

prior to embarking on your trip in order to give him or her your itinerary. Generally, your friends or mentor will offer to pick you up at the closest airport. If they are unavailable upon your arrival, it is recommended that you contact the registered Chinese student/scholar association by searching a local university's Web site. Give them your flight information so that they can better track your arrival time. A helpful hint during travel is to keep small denomination bills and coins to pay for phone calls and baggage cart rentals and for tipping luggage porters and taxi drivers.

3. Essential items to bring. It is very important that you bring notarized documents, including birth certificate, marriage certificate, and diplomas for identification, in order to obtain new documents and accounts in the U.S. and to facilitate the possible change of your visa status. If possible, browse public Web sites to obtain information on current weather conditions at your destination. You can get information from http://www.travel.com or from http://www.yahoo.com.

You can expect to find some appliances in your apartment, such as refrigerator, furnace, stove/oven, and water heater. Unless your apartment is advertised as furnished, you should not expect any furniture or bathroom or kitchen utensils to be provided. Bringing one or two cooking utensils will be helpful at the beginning of your stay. Bringing some simple bed linens and towels in your luggage is also advised. It may be necessary for you to sleep on the floor of your apartment (which may be carpeted) for a period until you can acquire a bed. Do not forget to bring any medicines, especially for the treatment of disorders in the digestive and respiratory systems. American drug stores sell many simple medications without a doctor's prescription, but they may be unfamiliar to you. Medications will be more important if your child travels with you. The international certificate of vaccination for your family should also be brought to America. It is an important document for the registration of your child at school in the U.S.

A Chinese-English Dictionary and an up-to-date textbook in your field are useful for reference. You may choose to bring some small gifts (not expensive) of native style to your mentor or colleague.

During your travel, you will be asked to fill out a form called I-94. Keep in mind that the I-94 form is even more important than your visa, which is just a permission to enter the U.S.: It determines how long you are authorized to stay in the U.S. It proves that you are in the country legally. The Immigration and Naturalization Service (INS) takes the I-94 back from you when you leave the U.S. When you come back from a third country, you are required to present your I-94 form at Customs.

When you first land in America, you will face many challenges, including language, new social customs, transportation, etc. Some important things to be done after your landing include the following.

AFTER ARRIVAL

1. *Registration.* As soon as possible, you should go to the administrative unit in your department or center for registration. The secretary there will ask for copies of your passport, visa, stamped I-94 card, and I-20 or IAP-66 form. You will fill out some important forms concerning wages, tax exemption, and health insurance. In addition, students or scholars from China should go to the Web site of the Chinese Embassy for online registration. The Web site is http://www.china-embassy.org.

   Keep in mind that the INS is enforcing an existing law, which requires that non-immigrants register their current address in the U.S. with the INS. Any change of address must be reported to the INS within 10 days of such change. Failure to register with the INS is a criminal offense. The Change of Address Card (Form AR-11) is available at the International Center and on the INS Web site at http://www.ins.usdoj.gov/graphics/formsfee/forms/ar-11.htm and should be completed upon arrival at your institution and at any time you change your address in the U.S.

2. *Apply for a Social Security Number (SSN).* SSN is the most important piece of identification you can get in America. It is given to you by the Social Security Administration. It is used by most government and non-government organizations to establish your identity. You will need a SSN to pay taxes, to find a job, to get a

credit card, or to get a loan. It is helpful to have a SSN if you want to open a bank account, get a driver's license, rent an apartment, or get your own phone number. Your credit history is associated with your SSN. Usually, the International Student/Scholar Center will give you a temporary SSN. Then you will receive a SSN card with your name and your SSN. Keep it in a safe location. You may get some relevant information through http://www.ssa.gov/online. Note that, effective July 2002, the Social Security Administration will verify documentation of all non-citizens with the INS before issuing a SSN. In addition to immigration documents, the Social Security Administration now requires additional proof of date of birth for those whose passports are less than a year old.

3. *Health insurance.* Participation in a health insurance program is required of international students or scholars by the U.S. government. For your benefit, health insurance could relieve the economic burden caused by the high cost of medical care. In most cases, your institutions will pay the cost of insuring you. You will then need to buy additional insurance for your family members. Browse the Web site http://www.student-resources.net or http://www.internationalstudentinsurance.com to find a proper insurance plan for you.

4. *Opening a bank account.* For your convenience and safety, you need to open a bank account promptly. You should not keep large amounts of cash with you. In many cases, your salary checks can automatically be deposited in your existing bank account on payday. You can usually check the balance of your account online.

5. *Obtaining telephone service.* There are different phone companies providing service in areas of the U.S. Consult the rental office to get relevant information. For long-distance calls (including international calls), you may consider buying phone cards at a savings in cost. Phone cards can also be ordered online, for instance, at http://www.chinamallonline.com .

6. *Obtaining a driver's license.* To obtain a driver's license, you will need to pass a written and a practical examination. The rules and

procedures vary in each state, but in general, these tests include a vision check, traffic sign recognition, understanding of traffic laws and safe driving practices, and an on-the-road test of your driving abilities. Again, some of this information is state-specific, and you should obtain the necessary study materials from the relevant government office upon arrival. Related information can be obtained at http://www.dmv.org. Before buying a used car, you should learn as much as you can about it, and you may consider having an independent mechanic examine it. Information on used cars or new cars can be viewed on the Internet at http://www.kbb.com or http://www.usedcars.com .

7. *Obtaining visa waivers.* Depending on circumstances, it may become necessary or desirable to extend your stay in the U.S. or to modify the conditions of your employment. Under such conditions, it will be necessary for you to apply to change your J visa. According to Section 212(e) of the Immigration and Nationality Act, J-1 visa holders must apply for a waiver of that requirement if they seek to remain in the U.S. beyond the end date of their programs or if they seek to submit an application to the Immigration and Naturalization Service for a change in visa status. The procedures for waiver application can be obtained from http://www.sino-education.org or http://www.travel.state.gov/jvw.htm. Keep in mind that laws and regulations pertaining to visas can change during your stay. It will be your responsibility to keep informed of these changes and to maintain a current address with the INS.

8. *Use the Internet.* As you are probably aware, the Internet is an excellent source of information on many subject matters pertaining to your travel and life in the U.S. Be aware that many Web sites change in content or address over time for a variety of reasons. If this happens, your best bet is to be familiar with the various search engines, such as http://www.google.com, http://www.netscape.com, http://www.lycos.com, or http://www.overture.com, which can help you find the information that you need.

The advice provided here is based on our own personal experience as visitors to the U.S. It is current at the time of the writing of this book, but you may have to seek updated information as Web sites change and procedures are modified. The authors sincerely hope that this information is helpful to you during your visit in the U.S.

Illustrated Laboratory Terminology

MEASURING 测定，測定

Volumes 体积，體積

1. Cordless pipetter
   直交流两用移液器
   直交流两用移液器

2. Pipette bulb
   吸管（移液管）吸耳球
   吸管（移液管）吸耳球

3. Pipettes
   吸管（移液管）
   吸管（移液管）

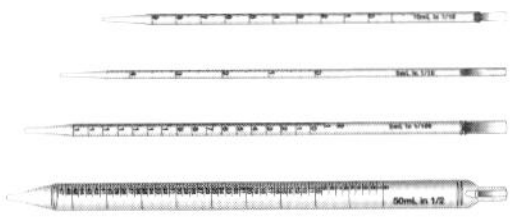

4. Pasteur pipette
   巴士德玻璃滴管（移液管）
   巴士德玻璃滴管（移液管）

5. Adjustable micropipetter
   可调式微量移液器
   可調式微量移液器

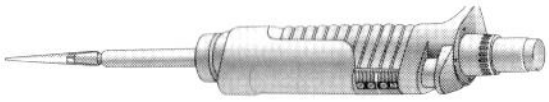

6. Micropipetter tips
   微量移液头
   微量移液頭

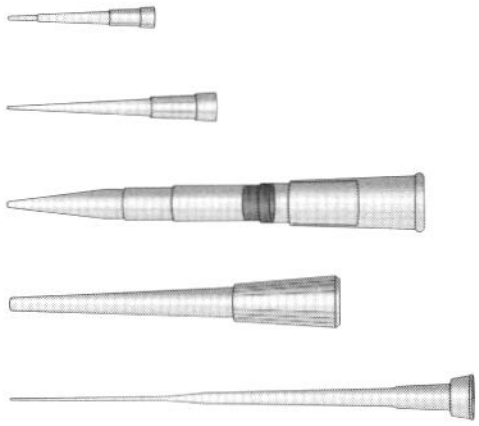

7. Multichannel pipetter

多孔道移液器

多孔道移液器

8. Repeating pipetter

连续移液器

連續移液器

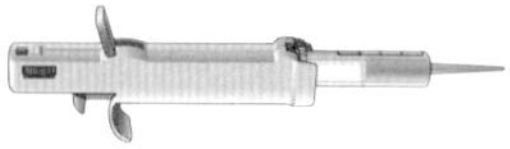

Optical Properties 光学器械，光學器械

9. Spectrophotometer

分光光度计

分光光度計

10. Microplate reader

微型板读数计

微型板讀數計

11. Cuvette

比色皿

比色皿

12. 384 well assay plate

384 孔试验板

384 孔試驗板

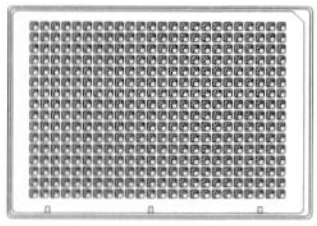

13a. 96 well assay plate opaque black

96 孔黑色不透明试验板

96孔黑色不透明試驗板

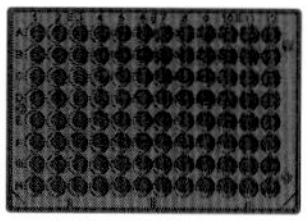

13b. 96 well assay plate opaque white

96 孔白色不透明试验板

96孔白色不透明試驗板

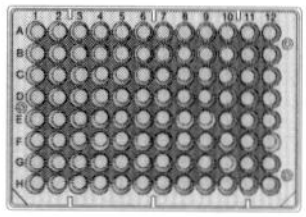

13c. 96 well assay plate clear

96 孔透明试验板

96 孔透明試驗板

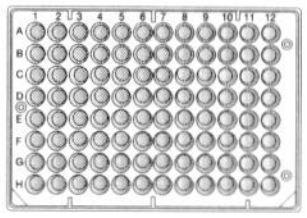

pH 酸碱度(pH), 酸鹼度

14. pH testing strips
酸碱度试纸
酸鹼度試紙

15. pH meter
pH 计
pH 計

Temperature 温度, 溫度

16. Thermometer
温度计
溫度計

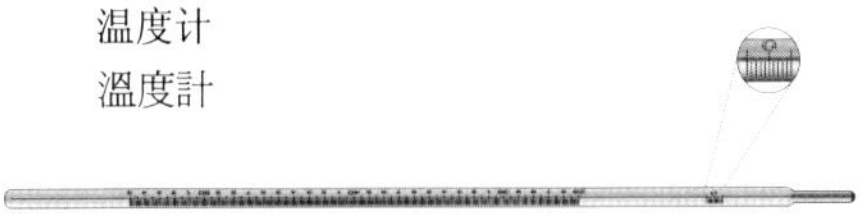

Mass 质量, 質量

17. Analytical balance
分析天平
分析天平

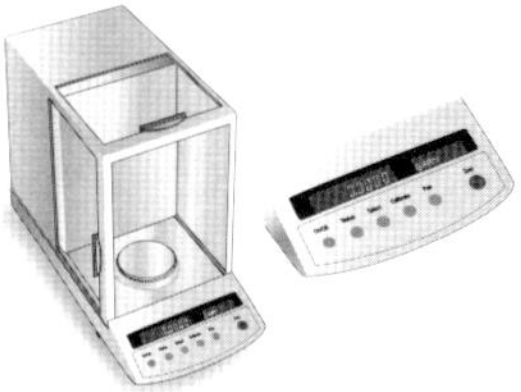

18. Reference weight set
标准砝码系列
標准砝碼系列

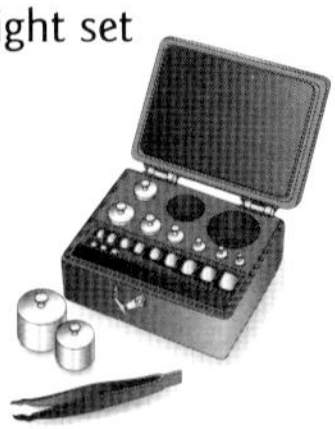

19. Disposable weighing dishes
一次性称重盘
一次性稱重盤

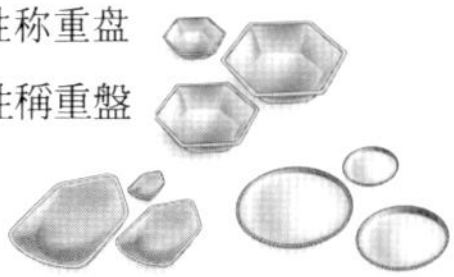

20. Spatula
药匙
藥匙

21. Antistatic brush
抗静电刷子
抗靜電刷子

Time 时间, 時間

22. Timer
    定时器
    定時器

Electrical Properties 电子器械, 電子器械

23. Voltmeter
    伏特计
    伏特計

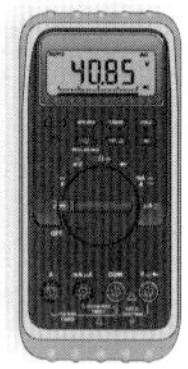

Dimensions 尺度, 尺度

24. Caliper
    测距尺
    測距尺

Radioactivity 放射性強度, 放射性強度

25. Scintillation counter
    闪烁计数仪
    閃爍計數儀

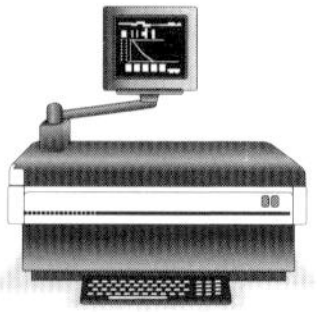

ADDITIONAL MEASURING TERMS

English	Simplified Chinese	Complex Chinese
absorbance	吸光率	吸光率
calibrate	校准	校準
capacity	容量，生产量，才能，能力，地位	容量，生產量，才能，能力，地位
Centigrade	摄氏度的，百分度的	攝氏度的，百分度的
centimeter	厘米	厘米
colorimeter	色度计，色量计	色度計，色量計
cut (v, n)	切，相交，削减，伤口	切，相交，削減，傷口
cuvette	小玻璃管，透明小容器，试管，比色皿	小玻璃管，透明小容器，試管，比色皿
extinction coefficient	消光系数	消光系數
Fahrenheit	华氏(温度)	華氏(溫度)
gallon	加仑	加侖
kilogram	千克	千克
liter	升	升
mass	大多数，质量，大量，聚集	大多數，群眾，大量，聚集
meter	米，公尺，计，表	米，公尺，計，儀表
metric	米制的，公制的	米制的，公制的
microbalance	微量天平	微量天平
microgram	微克	微克
micrometer	测微尺，千分尺，微米	測微尺，千分尺，微米
millennium	太平盛世，一千年	太平盛世，一千年
milligram	毫克	毫克
millimeter	毫米	毫米
millimole	毫摩尔	毫摩爾
million	百万	百萬
millionth	第一百万，百万分之一	第一百萬，百萬分之一
minimum	最小的，最低的，最小值	最小的，最低的，最小值

ADDITIONAL MEASURING TERMS

English	Simplified Chinese	Complex Chinese
molecular weight	分子量	分子量
nanogram	毫微克，纤克	毫微克，纖克
nanometer	十亿分之一(10^{-9}) 公尺， 毫微(10^{-9})米	十億分之一(10^{-9})公尺，毫微(10^{-9})米
nanomole	毫微(10^{-9})摩尔，毫微摩	毫微(10^{-9})摩爾，毫微摩
osmolality	重量摩尔渗透压浓度	重量摩爾滲透壓濃度
osmolarity	摩尔渗透压浓度	摩爾滲透壓濃度
permanent marker	永久标记笔	永久標記筆
pH buffer	pH 缓冲液	pH緩沖液
pH paper	pH 试纸	pH試紙
pint	品脱	品脫
pipette	移液管	移液管
pipette tip	移液管套头，枪头	移液管套頭，槍頭
pipettor	移液器	移液器
pound	磅	磅
quantitative	数量的，定量的	數量的，定量的
quart	夸脱	夸脫
ruler	统治者，管理者，直尺，划线板	統治者，管理者，直尺，劃線板
spectrophotometer	分光光度计	分光光度計
spectroscopy	光谱学，波谱学	光譜學，波譜學
temperature	温度	溫度
timer	定时器，计时员，记时员	定時器，計時員，記時員
UV	紫外线	紫外線
visible	可见的	可見的
volume	卷，册，体积，量，大量，音量	卷，冊，體積，量，大量，音量
wavelength	波长	波長

PROCESSING 加工处理设备，加工處理設備

Centrifuging 离心，離心

26. Microcentrifuge

微型离心机
微型离心机

27. Benchtop ultracentrifuge

台式超速离心机
台式超速离心机

28. Fixed angle rotor

固定式角转子
固定式角轉子

29. Swinging bucket rotor

摆动式桶状旋转子
擺動式桶狀旋轉子

30. Full size ultracentrifuge

大型超速离心机
大型超速離心機

31. Snap cap tube

具盖试管
具蓋試管

32. Plug cap tube

具塞试管
具塞試管

33. Round bottom culture tube

圆底培养管
圓底培養管

34. Conical bottom culture tube

锥底试管
錐底培養管

35. Cryogenic vials

低温瓶
低溫瓶

36. Microcentrifuge tubes

微型离心管
微型离心管

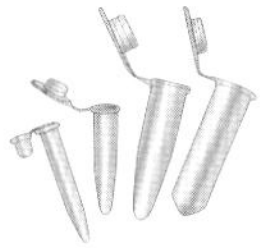

Cooling 冷藏设备，冷藏設備

37. Ice bucket
    冰桶
    冰桶

38. Refrigerating circulator
    冷藏循环器
    冷藏循環器

39. Liquid nitrogen storage tank
    液氮储存罐
    液氮儲存罐

Heating 加热设备，加熱設備

40. Heating bath
    加热水浴
    加熱水浴

41. Microwave oven
    微波炉
    微波爐

42. Oven
    烤箱
    烤箱

43. Hot plate
    电炉板
    電爐板

44. Heating block
    具孔加热板
    具孔加熱板

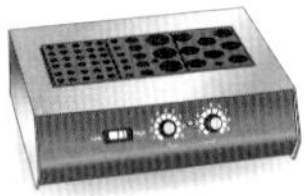

45. Thermomixer
    振荡加热器
    振蕩加熱器

46. Heat gun
    热风干燥器
    熱風干燥器

Homogenizing 匀浆设备，勻漿設備

47. Tissue grinder
组织匀浆器
組織勻漿器

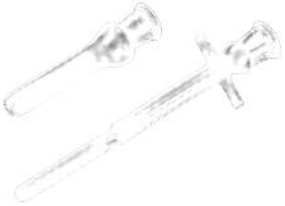

48. Sonicating probe
超声匀浆器
組織勻漿器

Mixing 混匀设备，混勻設備

49. Rocking platform
摇床
搖床

50. Tumbling mixer
翻转混匀器
翻轉混勻器

51. Vortex mixer
涡旋混匀器
渦旋混勻器

52. Sonicating bath
超声水浴
超聲水浴

53. Stir plate
搅拌台
攪拌台

54. Stir bars
搅拌转子
攪拌轉子

Storing 储存设备，儲存設備

55. Refrigerator/freezer
冷藏/冷冻冰箱
冷藏/冷凍冰箱

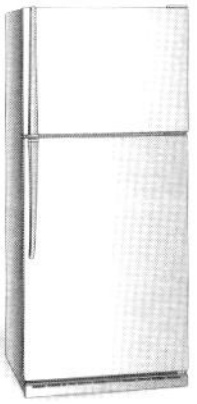

56. Ultra freezer
超低温冰箱
超低溫冰箱

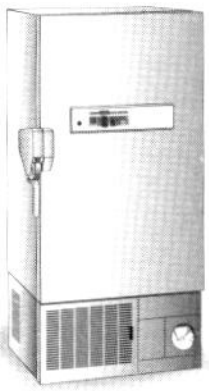

57. Freezer rack
冷冻架
冷凍架

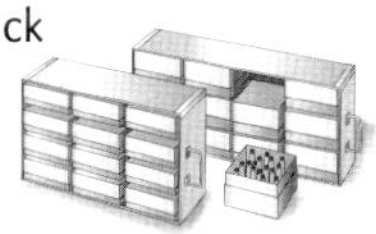

Drying 干燥设备，干燥設備

58. Dessiccator
干燥皿
干燥皿

59. Gel dryer
凝胶干燥器
凝膠干燥器

60. Lyophilizer
冻干器
凍干器

Filtering 过滤装置，過濾裝置

61. Sterile filtration unit
消毒过滤器
消毒過濾器

62. Vacuum filtration unit
负压过滤器
負壓過濾器

63. In-line air filter
过滤嘴
空氣過濾嘴

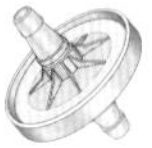

ADDITIONAL PROCESSING TERMS

English	Simplified Chinese	Complex Chinese
agitate	搅动，激动	攪動，搖動，煽動，激動
aliquot	能整除，分装	能整除，部分的，分裝
centrifuge	离心机	離心機
centrifuge tube	离心管	離心管
clamp	夹子，夹具，夹紧	夾子，夾具，夾緊
compress (v)	压缩，摘要叙述	壓縮，摘要敘述
compressor	压缩物，压缩机，收缩肌	壓縮物，壓縮機，收縮肌
condense	浓缩，精简	濃縮，精簡
condenser	冷凝器，电容器	冷凝器，電容器
cool (adj)	凉爽，冷静的	涼爽，冷靜的
cool (v)	(使)冷却	(使)冷卻
copy (n, v)	副本，拷贝，复印，复制，抄袭	副本，拷貝，複印，複製，抄襲
decelerate	减速	減速
decompress	减压，解压	減壓，解壓
deflate	放气，抽出空气，使缩小，紧缩	放氣，抽出空氣，使縮小，緊縮
degrade	降解，堕落，退化	降解，墮落，退化
dehydrate	脱水	脫水
denature	变性，改变本性	變性，改變本性
disperse	分散，散开，疏散	分散，散開，疏散
dissociate	分离，游离，分裂	分離，游離，分裂
distilled	由蒸馏得来的	由蒸餾得來的
distribute	分发，分配，散布，分布	分發，分配，散布，分布
divide	分，划分，分开，隔开	分，劃分，分開，隔開
dry	干的，干燥的，口渴的，干燥	乾的，乾燥的，口渴的，乾燥
dry ice	干冰	乾冰
dryness	干，干燥	乾，乾燥

ADDITIONAL PROCESSING TERMS

English	Simplified Chinese	Complex Chinese
excision	切除，删除，切除术	切除，刪除，切除術
extract (v)	榨，摘录，开方，析取，吸取	榨，摘錄，開方，析取，吸取
fixed angle	固定角度	固定角度
fractionate	使分馏	使分餾
freeze	结冰，冷冻，冻结	結冰，冷凍，凍結
freezer	冷冻室，冷藏工人	冷凍室，冷藏工人
g-forces	重力	重力
heat (n)	热，热烈，高潮，压力	熱，熱烈，高潮，壓力
heat (v)	加热，使激动，发怒	加熱，使激動，發怒
heater	加热器，发热器	加熱器，發熱器
heating block	加热板	加熱板
homogenize	均质化，使均匀	均質化，使均勻
homogenous	同质的，纯系的	同質的，純系的
ignite	点火，点燃	點火，點燃
immobilize	固定不动	固定不動
incline	倾斜，斜坡，倾向	傾斜，斜坡，傾向
inject	注射，注入	注射，注入
instill	灌注	灌注
knife	刀，餐刀	刀，餐刀
liquid nitrogen	液氮	液氮
lyophilize	使冻干	使凍乾
lysate	细胞裂解物	細胞裂解物
lyse	裂解	裂解
magnifier	放大镜，放大器	放大鏡，放大器
microcentrifuge	微量离心机	微量離心機
microfuge	微量离心	微量離心
mince	切碎	切碎

ADDITIONAL PROCESSING TERMS

English	Simplified Chinese	Complex Chinese
mixer	混合器，搅拌器	混合器，攪拌器
orbital rocker	旋转混匀器	旋轉混勻器
pellet (n)	小球，沉淀物	小球，沉澱物
permeable	有浸透性的，能透过的	有浸透性的，能透過的
process (n)	过程，作用，方法，步骤，处理	過程，作用，方法，步驟，處理
processor	处理器，处理者	處理器，處理者
pump (n, v)	泵，抽水机，抽水，抽吸	泵，抽水機，抽水，抽吸
radius	半径，范围，界限	半徑，範圍，界限
razor blade	剃刀	剃刀
recover	恢复，痊愈，复原	恢復，痊癒，復原
refrigerate	使冷却，使清凉，冷藏	使冷卻，使清涼，冷藏
renature	复性，使变性蛋白恢复	復性，使變性蛋白恢復
rocker	混摇器	混搖器
rotor	转子	轉子
RPM	每分钟转数	每分鐘轉數
rubber policeman	橡皮细胞刮脱器	橡皮細胞刮脫器
scrape	刮，擦，刮掉	刮，擦，刮掉
scraper	刮刀，刮器	刮刀，刮器
section (n)	部分，断片，部件，节，项，页，地域	部分，斷片，部件，節，項，頁，地域
section (v)	切片	切片
shake (v)	摇动，摇，颤抖，震动	搖動，搖，顫抖，震動
shaker	摇动器，混合器	搖動器，混合器
slice (n)	薄片，切片，部分	薄片，切片，部分
slice (v)	切	切
slippery	滑的，光滑的	滑的，光滑的
sonicator	超声波仪	超聲波儀

ADDITIONAL PROCESSING TERMS

English	Simplified Chinese	Complex Chinese
spatula	压舌板，药匙，刮铲	壓舌板，藥匙，刮鏟
steam (n)	蒸汽，水汽，蒸发，蒸	蒸汽，水汽，蒸發，蒸
stir	搅动，搅和	攪動，攪和
sublimate	升华，纯化	升華，純化
swinging bucket	摇动水桶	搖動水桶
tachometer	转速计，流速计	轉速計，流速計
tachymeter	准距仪	準距儀
test (v)	测验，试验，检验	測驗，試驗，檢驗
ultra freezer	超低温冰箱	超低溫冰箱
ultrabalance	微量天平	微量天平
vacuum (n, adj)	真空，真空的	真空，真空的
vacuum trap	负压吸引瓶	負壓吸引瓶
vortex mixer	旋涡混匀	旋渦混勻
warm	温暖的，兴奋的，保暖物	溫暖的，興奮的，保暖物

CULTURING 培养设备, 培養設備

64. Culture hood

培养台
培養工作台

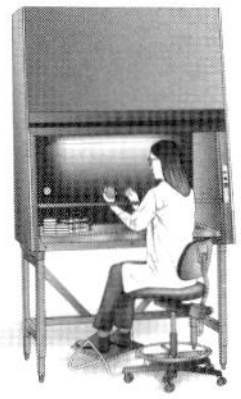

65. Culture flask

培养瓶
培養瓶

66. Culture dishes

培养皿
培養皿

67. Culture media

培养液
培養液

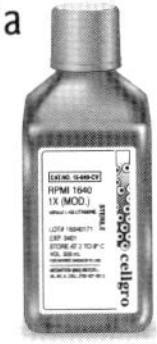

68. Cell scraper

细胞刮铲
細胞刮鏟

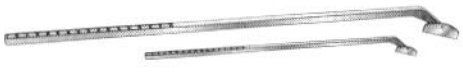

69. Bunsen burner

本生灯(煤气灯)
本生燈(煤气燈)

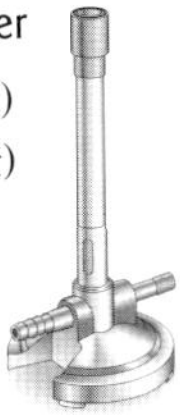

70. Vacuum trap

负压吸引瓶
負壓吸引瓶

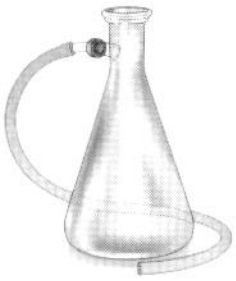

71. Incubator

培养箱
培養箱

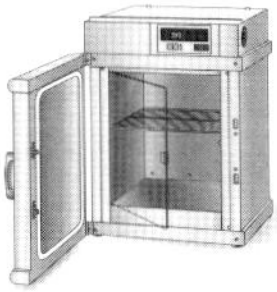

72. Culture vessel

培养容器
培養瓶

73. Loop

细菌接种环
細菌接种環

74. Culture shaker

培养摇床
培養搖床

ADDITIONAL CULTURING TERMS

English	Simplified Chinese	Complex Chinese
12-well plate	12 孔培养板	12 孔培養板
24-well plate	24 孔培养板	24 孔培養板
48-well plate	48 孔培养板	48 孔培養板
6-well plate	6 孔培养板	6 孔培養板
96-well plate	96 孔培养板	96 孔培養板
agarose	琼脂糖	瓊脂糖
cell culture	细胞培养	細胞培養
cell plate	细胞培养板	細胞培養板
cellular	细胞的	細胞的
clone (n)	无性系，无性繁殖，克隆	無性系，無性繁殖，克隆
clone (v)	无性繁殖，复制	無性繁殖，複製
colony	殖民地，侨民，菌落	殖民地，僑民，菌落
deprive	剥夺，丧失	剝奪，喪失
dish	盘，碟，提出	盤，碟，提出
dominant	有统治权的，支配的，显性的	有統治權的，支配的，顯性的
doubling time	双倍时间	雙倍時間
endotoxin	内毒素	內毒素
epithelium	上皮，上皮细胞	上皮，上皮細胞
expand	使膨胀，详述，扩张，发展	使膨脹，詳述，擴張，發展
fetal calf serum	胎牛血清	胎牛血清
filter (n)	滤波器，过滤器，滤光器	濾波器，過濾器，濾光器
filter (v)	过滤，渗透，用过滤法除去	過濾，滲透，用過濾法除去
flask	培养瓶	培養瓶
gram negative	革兰氏阴性	革蘭氏陰性
gram positive	革兰氏阳性	革蘭氏陽性
growth	生长，发育，栽培，种植	生長，發育，栽培，種植
growth factor	生长因子	生長指數

ADDITIONAL CULTURING TERMS

English	Simplified Chinese	Complex Chinese
growth media	培养基	生長培養基
Hanks' buffered saline	Hank 氏缓冲盐水	Hank氏緩沖鹽水
haploid	（染色体）单倍体，单一的	（染色體）單倍體，單一的
humidity	湿度，潮湿	濕度，潮濕
hybrid	混合物，杂交的	混合物，雜交的
hydrate (n)	氢氧化物	氫氧化物
hydrocortisone	氢化可的松	氫化考地松
immortalize	使不灭，使不朽，不老	使不滅，使不朽，不老
in situ	原位	原位
incubation	孵育，熟虑	孵育，熟慮
incubator	培养箱	培養箱
infection	感染，传染，影响，传染病	感染，傳染，影響，傳染病
medium (n)	介质，培养液	介質，培養液
medium (adj)	中间的，中等的，半生熟的	中間的，中等的，半生熟的
microbial	微生物的，由细菌引起的	微生物的，由細菌引起的
mycoplasma	支原体，支原菌	支原體，支原菌
permeability	渗透性	滲透性
permeabilize	使渗透	滲透
phosphate buffered saline	磷酸盐缓冲液	磷酸鹽緩沖液
phospholipase	磷脂酶	磷脂酶
plaque	噬菌斑	噬菌斑
plasmid	质粒，质体	質粒，質體
polycarbonate	聚碳酸酯	聚碳酸酯
polypropylene	聚丙烯	聚丙烯
polystyrene	聚苯乙烯	聚苯乙烯
pore	毛孔，小孔，气孔	毛孔，小孔，氣孔

ADDITIONAL CULTURING TERMS

English	Simplified Chinese	Complex Chinese
positive	肯定的，积极的，正的，阳的	肯定的，積極的，正的，陽的
pre-incubate	预先孵育	預先孵育
rescue (v)	救援，营救	救援，營救
resistance	反抗，抵抗，阻力，电阻，阻抗	反抗，抵抗，阻力，電阻，阻抗
retrovirus	逆转录酶病毒	逆轉錄酶病毒
satellite	人造卫星	人造衛星
serum	血清，免疫血清	血清，免疫血清
starve	(细胞)乏生长因子培养	(細胞)乏生長因子培養
stress (n, v)	重压，压力，强调，重读	重壓，壓力，強調，重讀
supplement	补充，附录，增刊	補充，附錄，增刊
T150 flask	T150 培养瓶	T150 培養瓶
T25 flask	T25 培养瓶	T25培養瓶
T75 flask	T75 培养瓶	T75培養瓶
tissue	组织	組織
trypsin	胰蛋白酶	胰蛋白酶
vector	载体，标量	載體，標量
viability	生存能力，发育能力	生存能力，發育能力
viable	能养活的，能生育的，可行的	能養活的，能生育的，可行的
virus	病毒，毒害，恶毒	病毒，毒害，惡毒

DOCUMENTING 文件制作，文件製作

75. Laboratory notebook

试验记录本
試驗記錄本

76. Gel documentation system

凝胶成象系统
凝膠成象系統

77. UV/visible transilluminator

紫外光/可见光透照仪
紫外光/可見光透照儀

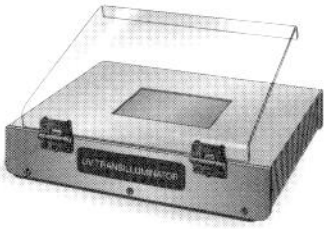

78. Strip chart recorder

疑胶梳
疑膠梳

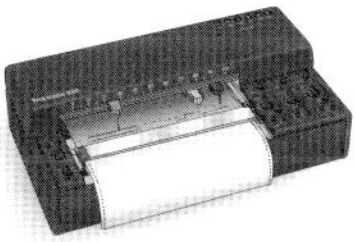

ADDITIONAL DOCUMENTING TERMS

English	Simplified Chinese	Complex Chinese
2-way ANOVA	双向方差分析	雙向方差分析
abstract	抽提，提取，摘要	抽提，提取，摘要
copyright	版权，著作权	版權，著作權
diskette	磁盘，磁碟	磁盤，磁碟
ink	墨水	墨水
inkjet	喷墨	噴墨
journal	定期刊物，杂志	定期刊物，雜誌
laptop computer	手提电脑	手提電腦
line graph	线图	線圖
manuscript	手稿，原稿，草稿	手稿，原稿，草稿
monitor	监测，监控，监听器	監測，監控，監聽器
notebook	笔记本	筆記本
paper	纸，报纸，文件，论文，用纸包装	紙，報紙，文件，論文，用紙包裝
parametric	参数的，参变量的	參數的，參變量的
percent	百分比，百分数	百分比，百分數
pie chart	圆图	圓圖
post test	后续检验	後續檢驗
printer	打印机，印刷工	打印機，印刷工
printout	打印输出	打印輸出
publication	出版物，出版，发行，公布，发表	出版物，出版，發行，公布，發表
RAM	随机存取存储器，随机存取内存	隨機存取存儲器，隨機存取內存
reprint (n)	再版，复印，单行本	再版，複印, 單行本
review	回顾，复习，评论	回顧，復習，評論
review article	复习文献，综述	復習文獻, 綜述
scanner	扫描仪	掃描儀
scanning	扫描	掃描

ADDITIONAL DOCUMENTING TERMS

English	Simplified Chinese	Complex Chinese
scientific notation	科学符号	科學符號
sigmoidal	S 形曲线，S 形的	S形曲線，S形的
significance	意义，重要性	意義，重要性
standard	标准，规格，标准的，第一流的	標準，規格，標準的，第一流的
standard deviation	标准差	標準差
standard error	标准误	標準誤
statistical significance	统计学显著性意义	統計學顯著性意義
statistics	统计学，统计表	統計學，統計表
trademark	商标	商標
t-test	t-检验	t-檢驗

DETECTING & ASSAYING 测定和试验， 測試和試驗

Arrays 排列，排列

79. Microarray scanner

微列扫描仪
微列掃描儀

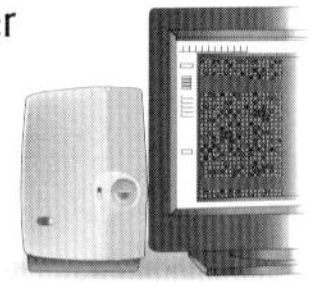

Blotting 印迹设备，印跡設備

80. Slot blot manifold

Slot 印迹
Slot 印跡

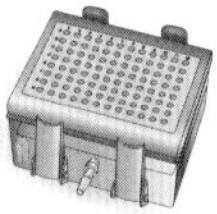

81. Blotting paper

印迹用滤纸
印跡用濾紙

82. Blotting membranes

印迹膜
印跡膜

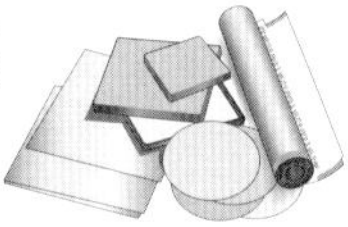

Microscopy 显微镜， 顯微鏡

83. Stereo dissecting microscope

立体解剖显微镜
立體解剖顯微鏡

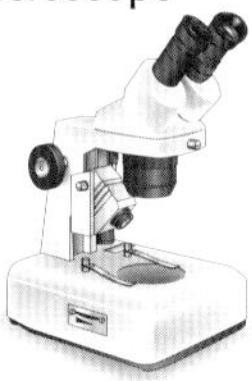

84. Inverted microscope

倒置显微镜
倒置顯微鏡

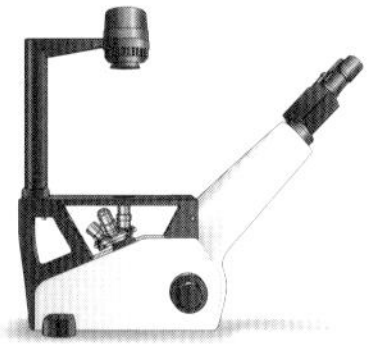

85. Upright microscope

直立显微镜
直立顯微鏡

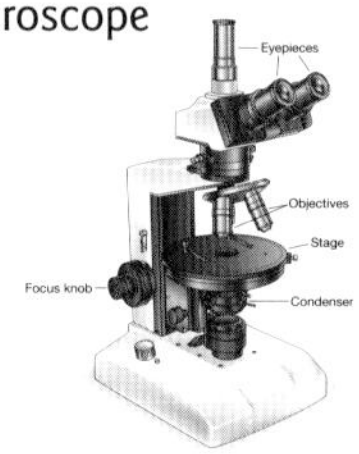

86. Microscope slides

显微镜用载玻片
顯微鏡用載玻片

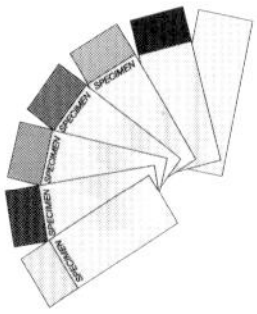

PCR 聚合酶链反应仪，聚合脢鏈反應儀

87. PCR thermocycler

PCR 扩增仪

PCR 擴增儀

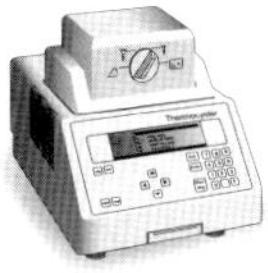

88. PCR plate

PCR 板

PCR 板

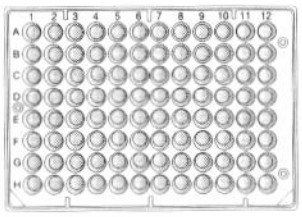

ADDITIONAL DETECTING & ASSAYING TERMS

English	Simplified Chinese	Complex Chinese
antibody	抗体	抗體
aperture	孔，缝隙，光圈，孔径	孔，縫隙，光圈，孔徑
band	条带，队，乐队，联合，结合	條帶，隊，樂隊，聯合，結合
bifocal	双焦点的	雙焦點的
chemiluminescence	化学光，化学发光	化學光，化學發光
coverslip	盖玻片	蓋玻片
destain	(使)标本褪色	標本褪色
detector	发现者，侦察器，检波器，检电器	發現者，偵察器，檢波器，檢電器
dial	拨号，打电话	撥號，打電話
electrode	电极	電極
electron microscope	电子显微镜	電子顯微鏡
electrophoresis	电泳，现象	電泳，現象
enzyme linked immunoassay (ELIZA)	酶联免疫测定	酶聯免疫測定
expose	使暴露，受到，使曝光，揭露	使暴露，受到，使曝光，揭露
horseradish peroxidase	辣根过氧化物酶	辣根過氧化物酶
Hz	赫兹	赫茲
ice	冰，结冰，冰冻	冰，結冰，冰凍
immunoprecipitation	免疫沉淀	免疫沉澱
infrared	红外线的，红外的	紅外線的，紅外的
integrate	使成整体，使一体化，结合	使成整體，使一體化，結合
isoelectric point	等电离点	等電離點
luminescence	发光	發光
luminol	发光氨	發光氨

ADDITIONAL DETECTING & ASSAYING TERMS

English	Simplified Chinese	Complex Chinese
mass spectrometer	质谱仪	質譜儀
maximum	最多量，最大限度，极大	最多量，最大限度，極大
microscope	显微镜	顯微鏡
mobility	活动性，灵活性，机动性	活動性，靈活性，機動性
nitrocellulose	硝基纤维素膜	硝基纖維素膜
Northern (RNA) blot	Northern RNA 印迹	Northern RNA 印跡
PCR	多聚酶链反应	多聚酶鏈反應
peroxidase	过氧化物酶	過氧化物酶
polymerase chain reaction	PCR，聚合酶链反应	PCR，聚合脢鏈反應
primary	第一	第一
protein A	A 蛋白	A蛋白質
protein G	G 蛋白	G蛋白質
PVDF	聚偏二氟乙烯	聚偏二氟乙烯
radioimmunoassay	放射免疫测定	放射免疫測定
response	回答，响应，反应	回答，響應，反應
signal (n, adj)	信号，信号的	信號，信號的
signal (v)	发信号，用信号通知	發信號，用信號通知
Southern (DNA) blot	Southern DNA 印迹	Southern DNA 印跡
stage	发展的进程，阶段或时期	發展的進程，階段或時期
Western (protein) blot	蛋白印迹	蛋白印跡

SEPARATING & FRACTIONATING 分离和分馏设备，分离和分餾設備

Chromatography 色谱仪，色谱仪

89. FPLC system
液相色谱系统
液相色譜系統

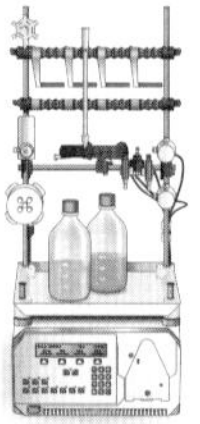

90. Chromatography columns
层析柱(色谱柱)
層析柱(色譜柱)

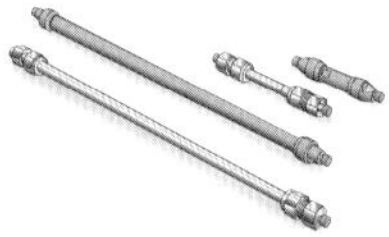

91. Chromatogram
层析图(色谱图)
層析圖(色譜圖)

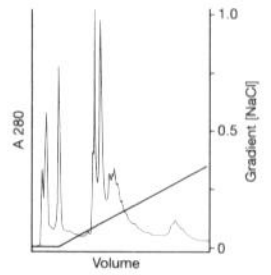

Electrophoresis 电泳设备，電泳設備

92. Power supply
电源
電源

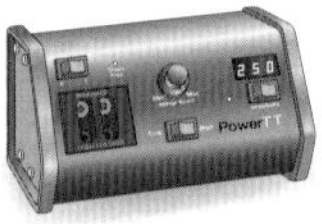

93. Transfer apparatus
电转仪
電轉儀

94. Horizontal gel electrophoresis system
水平凝胶电泳仪
水平凝膠電泳儀

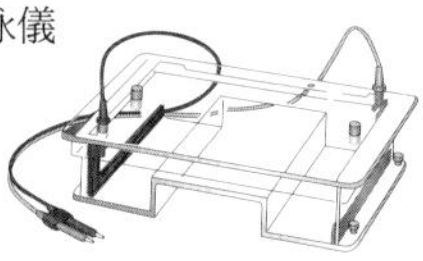

95. Vertical electrophoresis system
垂直电泳仪
垂直電泳儀

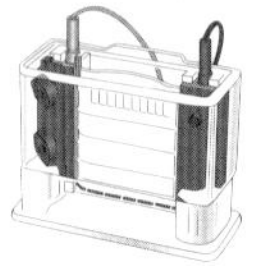

96. Mini gel system
迷你凝胶电泳仪
小型凝膠系統

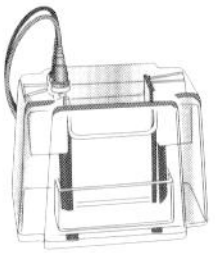

97. Sequencing gel electrophoresis system
测序凝胶电泳系统
測序凝膠電泳系統

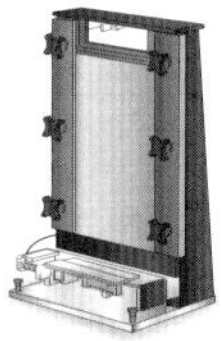

98. Gel combs

凝胶梳
凝膠梳

99. Pre-cast gel ready

预灌凝胶板
預灌凝膠板

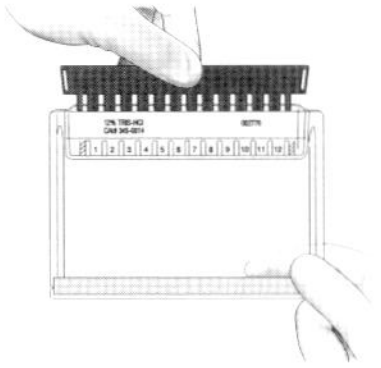

100. Electrodes

电极
電極

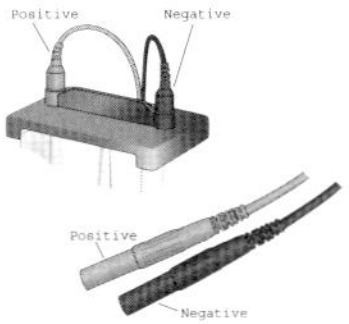

101. Fraction collector

组份收集器
組份收集器

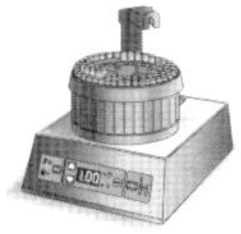

102. Autosampler vial

自动集样管
自動集樣管

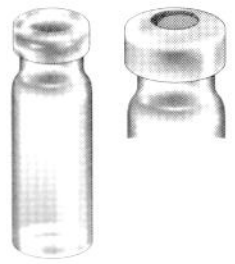

ADDITIONAL SEPARATING & FRACTIONATING TERMS

English	Simplified Chinese	Complex Chinese
chromatography	层析	層析
HPLC	高效液相色谱	高效液相色譜
hydrophilic	亲水的，吸水的	親水的，吸水的
hydrophobic	狂犬病的，疏水性	狂犬病的，恐水症的
isocratic	权利平等主义的	權利平等主義的
power supply	电源	電源
resolution	坚定，决心，决定，决议	堅定，決心，決定，決議
resolve	决心，解决，溶解	決心，解決，溶解
SDS PAGE	十二烷基磺酸钠聚丙烯酰胺凝胶电泳	十二烷基磺酸鈉聚丙烯酰胺凝膠電泳
thin layer chromatography	薄层色谱法	薄層色譜法
Volt	伏特	伏特
voltage	电压，伏特数	電壓，伏特數

SUPPLIES & EQUIPMENT 易耗品及设备, 易耗品及設備

103. Lab stool

实验凳
實驗凳

104. Laser pointer

激光手电
激光手電

105. Calculator

计算器
計算器

106. Electrical receptacle

电源插座
電源插座

107. Outlet strip

电源插座板
電源插座板

108. Glass stoppers

玻璃瓶塞
玻璃瓶塞

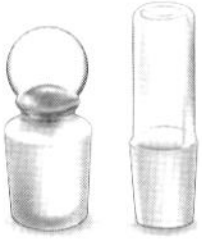

109. Rubber stopper

橡皮瓶塞
橡皮瓶塞

110. Funnel

漏斗
漏斗

111. Tray

托盘
托盤

112. Disposable syringe

一次性注射器
一次性注射器

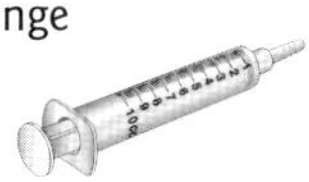

113. Gas-tight syringe

不透气注射器
不透气注射器

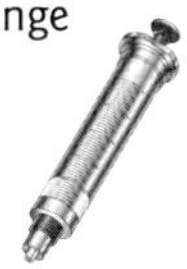

114. Microliter syringe

微量注射器
微量注射器

115. Syringe needle

注射用针头
注射用針頭

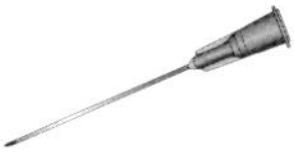

116. Scalpel

解剖刀
解剖刀

117. Razor blade

剃须刀片
剃須刀片

118. Scissors

剪刀
剪刀

119. Forceps

镊子
鑷子

120. Markers

记号笔
記號筆

121. Labeling tape

标记胶带
標記膠帶

122. Paraffin film

密封用石蜡薄膜
密封用石蜡薄膜

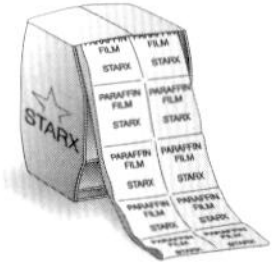

123. Vacuum pump

负压泵
負壓泵

124. Vacuum grease

负压泵密封膏
負壓泵密封膏

125. Tubing clamps

试管夹
試管夾

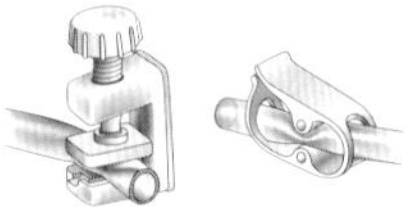

126. Tubing stopcock

试管旋塞阀
試管旋塞閥

127. Tubing connector

管道连接器
管道連接器

128. Connector

连接器
連接器

129. Tubing

管形材料
管形材料

130. Flowmeter

流量计
流量計

131. Gas cylinder support

气罐固定器
气罐固定器

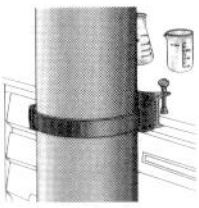

132. Gas tank regulator

气罐压力调节阀
气罐壓力調節閥

133. Squirt bottles

洗瓶
洗瓶

134. Bottle brush

瓶刷
瓶刷

135. Dishwashing soap

洗刷用肥皂
洗刷用肥皂

136. Lab sink

实验室水槽
實驗室水槽

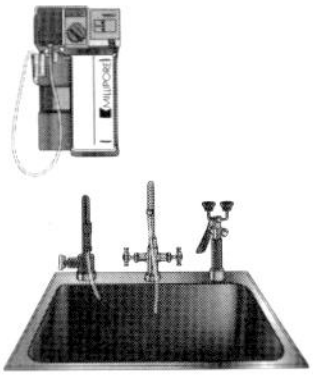

137. Ice scoop

冰勺
冰勺

138. Ice machine

制冰机
制冰機

ADDITIONAL SUPPLIES & EQUIPMENT

English	Simplified Chinese	Complex Chinese
adhesive	带粘性的，粘合剂	帶粘性的，膠粘，粘合劑
battery	电池	電池
CD	光碟	光碟
chair	教授职位，主席位，要职，椅子	教授職位，主席位，要職，椅子
foil	箔，金属薄片，衬托	箔，金屬薄片，襯托
forceps	镊子，钳子	鑷子，鉗子
glue	胶，胶水，粘合，粘贴	膠，膠水，粘合，粘貼
label (n, v)	标签，贴标签	標簽，貼標簽
marker	标记，标志，记号	標記，標志，記號
scissors	剪刀	剪刀
tape (n, v)	带子，录音带，磁带，录音	帶子，錄音帶，磁帶，錄音
tweezers	镊子	鑷子

SAFETY 安全用品和装置，安全用品和裝置

Apparel 服饰，服飾

139. Safety glasses
安全眼镜
安全眼鏡

140. Safety goggles
护目镜
護目鏡

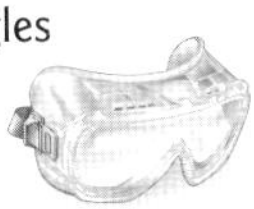

141. Laboratory coat
工作服
工作服

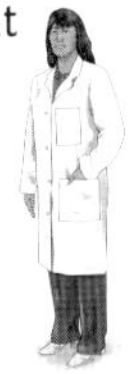

142. Shoe covers
脚套
腳套

143. Respirator
过滤式面罩
過濾式面罩

144. Dust mask
防尘面罩
防塵面罩

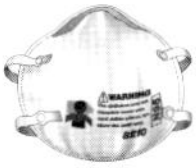

145. Ear muff
耳套
耳套

146. Ear plugs
耳塞
耳塞

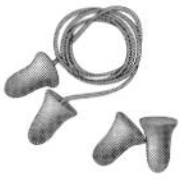

147. Latex glove
乳胶手套
乳膠手套

148. Rubber gloves
橡皮手套
橡皮手套

149. Cryogenic gloves
低温用手套（棉手套）
低溫用手套（棉手套）

150. Hand protector
护掌套
護掌套

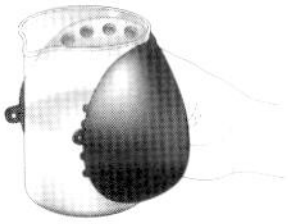

Equipment 设备，設備

151. Fume hood
通风用烟罩
通風用煙罩

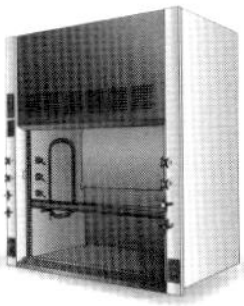

152. Radiation monitor
辐射监测器
輻射監測器

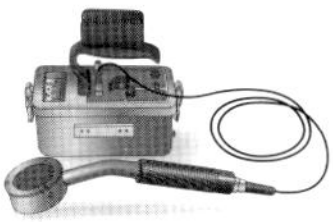

153. Beta emitter radiation shield

β辐射防护板

β輻射防護板

154. Beta storage container

β輻射儲存盒

β輻射儲存盒

155. Fire extinguisher

灭火器

滅火器

156. Hazard label

有害物标签

有害物標識

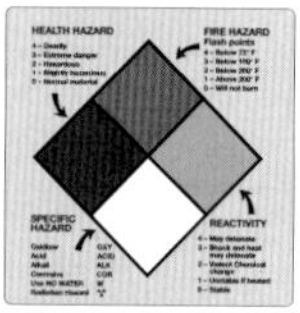

157. Flammable safety cabinet

易燃物安全柜

易燃物安全柜

158. First aid kit

急救箱

急救箱

159. Safety shower

应急淋浴器

應急淋浴器

160. Sink sprayer

水槽用喷水管

水槽用噴水管

161. Eyewash station

洗眼站

洗眼站

162. Spill containment kit

液体溢出处理盒

溢液處理盒

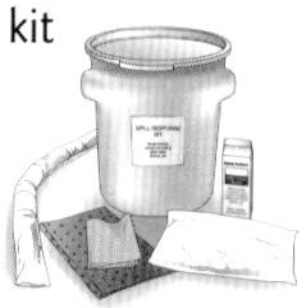

163. Biohazard container

生物有害物垃圾桶

生物有害物垃圾桶

164. Autoclave

高压灭菌器

高壓滅菌器

ADDITIONAL SAFETY TERMS

English	Simplified Chinese	Complex Chinese
^{125}I	碘 125	碘 125
^{32}P	磷 32	磷 32
^{33}P	磷 33	磷 33
^{3}H	氢 3 (氚)	氫 3 (氚)
alpha radiation	α 辐射, α 放射	α輻射，α放射，α發散
autoclave (n)	高压锅，高压灭菌器	高壓鍋，高壓滅菌器
beta radiation	β 辐射，β 放射，β 发光	β輻射，β放射，β發光
carcinogen	致癌物质	致癌物質
cardiotoxic	心脏毒的	心臟毒的
decay (v)	腐朽，腐烂，衰减，衰退	腐朽，腐爛，衰減，衰退
disinfect	消毒	消毒
explode	使爆炸，爆发，推翻，激发	使爆炸，爆發，推翻，激發
explosion	爆发，发出，爆炸	爆發，發出，爆炸
eye protection	眼保护	眼保護
eye wash station	洗眼台(站)，洗眼地点	洗眼站，洗眼地點
face mask	面罩	面罩
fire	火，火灾，射击，解雇	火，火災，射擊，解雇
floor mat	地板垫	地板墊
fume hood	排气罩	排氣罩
gamma radiation	γ 辐射，γ 放射	γ輻射，γ放射
gas	气体，煤气，毒气，汽油	氣體，煤氣，毒氣，汽油
gasket	垫圈，衬垫	墊圈，襯墊
gastrotoxic	胃毒的	胃毒的
genotoxic	遗传毒的	遺傳毒的
gloves	手套，戴手套	手套，戴手套
hearing	听力，倾听, 听证	聽力，傾聽, 聽證
heavy	重的，大量的	重的，大量的

ADDITIONAL SAFETY TERMS

English	Simplified Chinese	Complex Chinese
inflammation	炎症，发炎	炎症，發炎
inhale	吸入，吸气	吸入，吸氣
irradiate	照射	照射
lab coat	实验室工作服	實驗室工作服
mutagen	诱变剂	誘變劑
mutagenic	诱变的	誘變的
mutation	突变	突變
mycotoxin	霉菌毒菌(素)	霉菌毒菌(素)
myotoxic	肌毒的	肌毒的
nephrotoxic	肾脏毒的	腎臟毒的
neurotoxic	神经毒的	神經毒的
oxidant	氧化剂	氧化劑
oxidize	氧化	氧化
packaging	包装	包裝
pad (n)	垫，衬垫	墊，襯墊
pain	痛苦，劳苦，努力	痛苦，勞苦，努力
poison	毒药，毒害，中毒，放毒	毒藥，毒害，中毒，放毒
radiation badge	辐射剂量牌	輻射劑量牌
safe	安全的，可靠的，有把握的	安全的，可靠的，有把握的
safety glasses	安全眼镜	安全眼鏡
shield (n, v)	防护物，护罩，盾，盾状物，保护	防護物，護罩，盾，盾狀物，保護
shoe cover	鞋罩	鞋罩
shower (n)	淋浴，阵雨	淋浴，陣雨
smell	气味，嗅觉，臭味	氣味，嗅覺，臭味
snap cap	按式盖子	按式蓋子
software hardware	软件硬件	軟件硬件
sterile	消毒的，不育的	消毒的, 不育的

ADDITIONAL SAFETY TERMS

English	Simplified Chinese	Complex Chinese
sterile hood	消毒罩	消毒罩
sterilize	杀菌，消毒	殺菌，消毒
teratogen	致畸剂，畸胎剂	致畸劑，畸胎劑
teratogenic	致畸的	致畸的
toxic	有毒的	有毒的
toxin	毒素	毒素
virulent	剧毒的，致命的，恶性的	劇毒的，致命的，惡性的

Comprehensive Laboratory Terminology

English	Simplified Chinese	Complex Chinese
^{125}I	碘 125	碘 125
12-well plate	12 孔培养板	12 孔培養板
24-well plate	24 孔培养板	24 孔培養板
2-way ANOVA	双向方差分析	雙向方差分析
^{32}P	磷 32	磷 32
^{33}P	磷 33	磷 33
^{3}H	氢 3 (氚)	氫 3 (氚)
48-well plate	48 孔培养板	48 孔培養板
6-well plate	6 孔培养板	6 孔培養板
96-well plate	96 孔培养板	96 孔培養板
absorbance	吸光率	吸光率
abstract	抽提，提取，摘要	抽提，提取，摘要
accelerate	加速，促进	加速，促進
accord	一致，协定	一致，符合，調和，協定
acid	酸	酸
acidic	酸的，酸性的	酸的，酸性的
activate	刺激，激活	刺激，使活動，有活力

English	Simplified Chinese	Complex Chinese
acute	敏锐的，急性的	敏銳的，急性的，劇烈
adaptor	接合子，改编者，接口	接合子，改編者
add	增加，添加	增加，添加，加，加起來
adenovirus	腺病毒	腺病毒
adhesive	带粘性的，粘合剂	帶粘性的，膠粘，粘合劑
adjust	调整，调节	調整，調節，校准
adjustable	可调整的	可調整的
agarose	琼脂糖	瓊脂糖
age (n)	年龄，变老，时代	年齡，變老，時代
age (v)	变老，成熟	變老，成熟，上年紀
aggregate (n, v)	集合，聚集	合計，總計，集合，聚集
agitate	搅动，激动	攪動，搖動，煽動，激動
agonist	激动剂	激動劑
air	空气，通风	空氣，通風
alanine	丙氨酸	丙胺酸
alcohol	酒精，酒	酒精，酒
align	排列，使成一行	排列，使結盟，使成一行
alignment	队列，结盟	隊列，結盟
aliquot	能整除，分装	能整除，部分的，分裝
alkaline	碱，碱性的	鹼，鹼性的
allow	允许，承认	允許，承認
alpha radiation	α 辐射, α 放射	α輻射，α放射，α發散
amount (n)	数量	數量
amperage	安培数	安培數
ampere	安培	安培
amplicon	PCR 产物，PCR 结果	PCR 產物，PCR 結果
amplify	扩大，增强，扩大	擴大，增強，擴大

English	Simplified Chinese	Complex Chinese
anabolic	合成代谢的	合成代謝的
analyte	分析元素	分析元素
analytic	分析的，解析的	分析的，解析的
analytical	分析的，解析的	分析的，解析的
analyze	分析，分解	分析，分解
angle	角,角落，追逐	角，角落，追逐
angstrom	埃(长度单位)	埃(長度單位)
angular	有角的	有角的
animal	动物	動物
annual	每年的，年鉴	每年的，年鑒
ANOVA	方差分析	方差分析
antagonize	敌对，对抗	敵對，對抗
antenna	天线(复数)	天線(複數)
antibiotic	抗生素，抗生的	抗生素，抗生的
antibody	抗体	抗體
antioxidant	抗氧化剂，硬化防止剂	抗氧化劑，硬化防止劑
aperture	孔，缝隙，光圈，孔径	孔，縫隙，光圈，孔徑
apoptosis	凋亡	凋亡
apparatus	器械，设备，仪器	器械，設備，儀器
approve	赞成，批准，通过	贊成，滿意，通過
approximate (adj)	近似的，大约的	近似的，大約的
aquatic	水的，水上的，水生的	水的，水上的，水生的
aqueous	水的，水成的	水的，水成的
area	范围，面积，地区	範圍，面積，地區
arginine	精氨酸	精胺酸
arm	臂，武器，供给，装备	臂，武器，供給，裝備
arrange	安排，协商，排列	安排，協商，排列

English	Simplified Chinese	Complex Chinese
array	排列，编队，部署	排列，編隊，軍隊，
aseptic	防腐剂，无菌的	防腐劑，無菌的
asparagine	天冬酰胺	天門冬素，天門酸胺
aspartic acid	天冬氨酸	天(門)冬胺酸
aspirate	吸气	吸氣
aspiration	吸出, 热情，渴望	吸出, 熱情，渴望
assay	化验，检验，评价	化驗，檢驗，評價
associate (n, adj)	交往，合作人，副的	交往，合作人，副的
attach	系上，贴上，隶属于	系上，貼上，隸屬于
attachment	附件，附加装置，配属	附件，附加裝置，配屬
attack (v)	攻击，侵袭，抨击	攻擊，侵襲，抨擊
attempt	努力，尝试，企图	努力，嘗試，企圖
attenuate	削弱	削弱
attogram	10^{-18} 克	10^{-18} 克
attomole	10^{-18} 摩尔	10^{-18} 摩爾
atypical	非典型的	非典型的
audit (v)	审计，稽查，旁听	審計，稽查，旁聽
auditorium	听众席，观众席，礼堂	聽眾席，觀眾席，禮堂
autoclave (n)	高压锅，高压灭菌器	高壓鍋，高壓滅菌器
automate	使自动化，自动操作	使自動化，自動操作
automatic	自动的，机械的	自動的，機械的
autoradiography	放射线自显影术	放射線自顯影術
available	可得到的，有用的，有空的	可得到的，有用的，有空的
back	后面的，背部，后退，支持	後面的，背部，後退，支持
backcross	回交，回交杂种	回交，回交雜種
backward	向后地，退步，由好变坏	向後地，退步，由好變壞
bacterial	细菌的	細菌的

English	Simplified Chinese	Complex Chinese
bacteriophage	噬菌体	噬菌體
balance	秤，天平，平衡，对比	秤，天平，平衡，對比
ballistic	弹道学的，弹道的	彈道學的，彈道的
band	条带，队，乐队，联合，结合	條帶，隊，樂隊，聯合，結合
bar graph	条图	條圖
base	底部，基础，根据地，低级的	底部，基礎，根據地，低級的
basic	基本，要素，基础，碱性的	基本，要素，基礎，鹼性的
bath	沐浴，浴室	沐浴，浴室
bathe	沐浴，洗，洗澡	沐浴，洗，洗澡
battery	电池	電池
begin	开始，首先	開始，首先
bench	长椅子，试验台	長椅子，試驗台
beside	在旁边，和……比较	在旁邊，和……比較
best	最好的，最，极	最好的，最，極
beta radiation	β辐射，β放射，β发光	β輻射，β放射，β發光
better	较好的，更好的，更多	較好的，更好的
biannual	一年二次的	一年二次的
biennial	二年生植物，二年一次的	二年生植物，二年一次的
bifocal	双焦点的	雙焦點的
billion	十亿	十億
billionth	第十亿的，十亿分之一的	第十億的，十億分之一的
bind	绑，装订，结合	綁，裝訂，結合
binding	装订，捆绑，粘合物	裝訂，捆綁，粘合物
biochemical	生物化学的	生物化學的
bioinformatics	生物信息学	生物信息學
biological	生物学的	生物學的

English	Simplified Chinese	Complex Chinese
biology	生物学，生物	生物學，生物
birth	出生，产生，出身	出生，產生，出身
blend (v)	混合	混合
blind	瞎的，盲目的，缺乏眼光	瞎的，盲目的，缺乏眼光
block (n)	木块，街区，阻滞	木塊，街區，阻滯
block (v)	妨碍，阻塞	妨礙，阻塞
blood	血，血液，血统	血，血液，血統，
board	木板，甲板，部，上(船、飞机)	木板，甲板，部，上(船、飛機)
bolt	门闩，螺钉，跑掉	門閂，螺釘，跑掉
book (n)	书，卷，篇，名册，帐簿	書，卷，篇，名冊，帳簿
book (v)	登记，预定，控告	登記，預定，控告
book chapter	书籍章节	書籍章節
bottle	瓶子，用瓶装	瓶子，用瓶裝
bound	跳跃，范围，限度，限制	跳躍，範圍，限度，限制
branch (n)	分枝，分部，部门	分枝，分部，部門
break (v)	休息，暂停，破裂，削弱	休息，暫停，破裂，削弱
buffer (n)	缓冲液	緩沖液
bulb	灯泡	燈泡
butanol	丁醇	丁醇
byte	字节	字節
calibrate	校准	校準
cam	凸轮	凸輪
cap (n, v)	盖子，帽，盖在……顶上	蓋子，帽，蓋在……頂上
capacity	容量，生产量，才能，能力，地位	容量，生產量，才能，能力，地位
capillary	毛细管，毛状的，毛细作用的	毛細管，毛狀的，毛細作用的
capsid	衣壳，包围病毒的蛋白外壳	衣殼，包圍病毒的蛋白外殼

English	Simplified Chinese	Complex Chinese
carbohydrate	碳水化合物，糖类	碳水化合物，糖類
carcinogen	致癌物质	致癌物質
cardiotoxic	心脏毒的	心臟毒的
carrier	运送者，携带者	運送者，攜帶者
casing	包装，保护性的外套	包裝，保護性的外套
catabolic	分解代谢的，异化的	分解代謝的，異化的
catalyst	催化剂	催化劑
CD	光碟	光碟
ceiling	天花板，最高限度	天花板，最高限度
cell	单元，细胞，电池	單元，細胞，電池
cell culture	细胞培养	細胞培養
cell plate	细胞培养板	細胞培養板
cellular	细胞的	細胞的
centigrade	摄氏温度的，百分度的	攝氏溫度的，百分度的
centimeter	厘米	厘米
central	中央的，重要的	中心的，重要的
centrifuge	离心机	離心機
centrifuge tube	离心管	離心管
century	世纪，百年，百个	世紀，百年，百個
certificate	证书，发给证明书	證書，發給證明書
certify	证明，保证	證明，保證
chair	教授职位，主席位，要职，椅子	教授職位，主席位，要職，椅子
change (v)	改变，变化，转变	改變，變化，轉變，輔幣
change (n)	改变，变改，改造，兑换	改變，變改，改造，兌換
channel (n)	海峡，通道，沟，路线	海峽，通道，溝，路線
channel (v)	引导，开导	引導，開導
charcoal	木炭	木炭

English	Simplified Chinese	Complex Chinese
chemiluminescence	化学光，化学发光	化學光，化學發光
chemokine	趋化因子	化學導質
chill	寒战，寒心，扫兴，变冷	寒戰，寒心，掃興，變冷
chloroform	氯乙烷，氯仿	氯乙烷，氯仿
chop	砍，剁，商标	砍，剁，商標
chopper	断路器	斷路器
chromatin	染色质，核染质	染色質，核染質
chromatography	层析	層析
chromosome	染色体	染色體
chronograph	计时表，计秒表	計時表，計秒表
cilia	睫，纤毛	睫，纖毛
circuit board	电路板	電路板
circumference	圆周，周围	圓周，周圍
cis	顺式	順式
clamp	夹子，夹具，夹紧	夾子，夾具，夾緊
clean wipe	清扫擦	清掃擦
climate	气候	氣候
clip	夹子，回形针，剪短	夾子，回形針，剪短
clog (v)	障碍，阻塞	障礙，阻塞
clone (n)	无性系，无性繁殖，克隆	無性系，無性繁殖，克隆
clone (v)	无性繁殖，复制	無性繁殖，複製
coarse	粗糙的	粗糙的
codon	密码子	密碼子
coincidence	一致，相合，同时发生的事	一致，相合，同時發生的事
collateral	间接的	間接的
colony	殖民地，侨民，菌落	殖民地，僑民，菌落
colorimeter	色度计，色量计	色度計，色量計

English	Simplified Chinese	Complex Chinese
commercial	商业的，贸易的	商業的，貿易的
committee	委员会	委員會
compendium	纲要，概率	綱要，概率
complementary	补充的，互补的	補充的，互補的
complete (adj)	全部的，完全的，完成的，完成	全部的，完全的，完成的，完成
composite (n, adj)	合成物，复合的	合成物，複合的
compress (v)	压缩，摘要叙述	壓縮，摘要敘述
compressor	压缩物，压缩机，收缩肌	壓縮物，壓縮機，收縮肌
computer	计算机，电脑	計算機，電腦
computer program	计算机程序	計算機程序
concatenate (v)	连接	連接
concentration	集中，专心，浓缩，浓度	集中，專心，濃縮，濃度
condensate	冷凝物	冷凝物
condense	浓缩，精简	濃縮，精簡
condenser	冷凝器，电容器	冷凝器，電容器
conditional	有条件的	有條件的
confer	授予，赠与，协议，协商，交换意见	授予，贈與，協議，協商，交換意見
conference	会议，讨论会，协商会	會議，討論會，協商會
congress	代表大会，国会，议会	代表大會，國會，議會
conjugate (n)	结合, 同根词	結合, 同根詞
conjugate (v)	连接	連接
conjugation	结合	結合
connection	连接，关系，接线，线路，亲戚	連接，關系，接線，線路，親戚
conservative	保守的，守旧的，保守派	保守的，守舊的，保守派
conserve	保存，保藏	保存，包藏
constitutive	构成的，制定的	構成的，制定的

English	Simplified Chinese	Complex Chinese
construct (v)	建造，构造，创立	建造，構造，創立
container	容器，集装箱	容器，集裝箱
contract (n)	合同，契约，订约	合同，契約，訂約
contract (v)	使缩短，感染，订约	使縮短，感染，訂約
control	控制，支配，管理，调节，操作	控制，支配，管理，調節，操作
conventional	惯例的，常规的，习俗的，传统的	慣例的，常規的，習俗的，傳統的
converge	集中于一点，聚合，会聚	集中于一點，聚合，會聚
conversion	变换，转换	變換，轉換
cool (adj)	凉爽，冷静的	涼爽，冷靜的
cool (v)	(使)冷却	(使)冷卻
copious	很多的，广识的，丰富的	很多的，廣識的，豐富的
copy (n, v)	副本，拷贝，复印，复制，抄袭	副本，拷貝，複印，複製，抄襲
copyright	版权，著作权	版權，著作權
cordless	无绳的	無線的
core	核心，中心，果核	核心，中心，果核
corner	角落，绝境，把……难住，囤积	角落，絕境，把……難住，囤積
correlate (v)	使相互关联，和……相关	使相互關聯，和……相關
cosmid	装配型质粒	裝配型質粒
coupled	一对，连合，连接，结合	一對，連合，連接，結合
cover (n)	盖子，封面	蓋子，封面
cover (v)	覆盖，保护，包含，使用，涂	覆蓋，保護，包含，使用，塗
covering	遮盖物	遮蓋物
coverslip	盖玻片	蓋玻片
create	创造，创作，引起，造成	創造，創作，引起，造成

English	Simplified Chinese	Complex Chinese
creative	创造性的	創造性的
crossover	转线路，天桥，交叉	轉線路，天橋，交叉
cryptic	秘密的，含义模糊的，神秘的	秘密的，含義模糊的，神秘的
crystal	结晶状的，水晶，结晶，晶体	結晶狀的，水晶，結晶，晶體
cut (v, n)	切，相交，削减，伤口	切，相交，削減，傷口
cuvette	小玻璃管，透明小容器，试管，比色皿	小玻璃管，透明小容器，試管，比色皿
cycle (n)	周期，循环，轮转	周期，循環，輪轉
cyclin	细胞周期蛋白	細胞週期蛋白
cyclooxygenase	环加氧酶	環加氧酶
cylinder	圆筒，圆柱体，汽缸，气罐	圓筒，圓柱體，汽缸，氣罐
cysteine	半胱氨酸	半胱氨酸
cytokine	细胞因子	細胞因子
cytosine	胞嘧啶，氧氨嘧啶	胞嘧啶，氧氨嘧啶
cytotoxic	细胞毒的	細胞毒的
damage (v)	损伤，损害	損傷，損害
dark	黑，暗，无知，模糊	黑，暗，無知，模糊
darkness	黑暗，漆黑	黑暗，漆黑
deaf	聋的	聾的
decade	十年，十	十年，十
decay (v)	腐朽，腐烂，衰减，衰退	腐朽，腐爛，衰減，衰退
decelerate	减速	減速
decimal	十进的，小数的，十进制，小数	十進的，小數的，十進制，小數
decline (v)	下倾，下降，拒绝	下傾，下降，拒絕
decompress	减压，解压	減壓，解壓
defend	防护，辩护，方位	防護，辯護，方位

English	Simplified Chinese	Complex Chinese
defense	(美)国防部，防卫，抵御	(美)國防部，防衛，抵禦
deficiency	缺乏，不足	缺乏，不足
deficient	缺乏的，不足的，不完善的	缺乏的，不足的，不完善的
deflate	放气，抽出空气，使缩小，紧缩	放氣，抽出空氣，使縮小，緊縮
degenerate	退化的，退化	退化的，退化
degradation	降解，降格，退化	降解，降格，退化
degrade	降解，堕落，退化	降解，墮落，退化
dehydrate	脱水	脫水
deionized	去离子，除去离子	去離子，除去離子
delay	耽误，延迟，延期	耽誤，延遲，延期
delete	删除	刪除
demonstrate	示范，证明，论证，示威	示範，證明，論證，示威
denature	变性，改变本性	變性，改變本性
denominator	分母，命名者	分母，命名者
deposit (n)	堆积物，沉淀物，存款，押金	堆積物，沉澱物，存款，押金
deposit (v)	存放，堆积，沉淀	存放，堆積，沉澱
deprive	剥夺，丧失	剝奪，喪失
derivative	派生的，衍生的物质	派生的，衍生的物質
destain	(使)标本褪色	標本褪色
detector	发现者，侦察器，检波器，检电器	發現者，偵察器，檢波器，檢電器
detergent	清洁剂，去污剂	清潔劑，去污劑
determine	决心，测定	決心，測定
detoxify	使解毒	使解毒
development	进展，发展	進展，發展
dial	拨号，打电话	撥號，打電話
diatome	切片刀	切片刀

English	Simplified Chinese	Complex Chinese
dice	切成方块	切成方塊
differ	不同，不一致	不同，不一致
difference	差异，差别，分歧，争论	差異，差別，分歧，爭論
differentiate	区别，区分	區別，區分
direct	直接的，直率的，指引，指挥	直接的，直率的，指引，指揮
disallow	不接受	不接受
disapprove	不赞成	不贊成
discriminate	歧视，区别，区别待遇	歧視，區別，區別待遇
dish	盘，碟，提出	盤，碟，提出
disinfect	消毒	消毒
diskette	磁盘，磁碟	磁盤，磁碟
dismiss	解散，下课，开除	解散，下課，開除
dismutate	歧化	歧化
disparity	不一致，不同，不等	不一致，不同，不等
disperse	分散，散开，疏散	分散，散開，疏散
disprove	反驳，驳斥，证明……为误	反駁，駁斥，證明……為誤
dissociate	分离，游离，分裂	分離，游離，分裂
dissociation constant	分离常数	分離常數
distal	末梢的, 远端的	末梢的, 遠端的
distilled	由蒸馏得来的	由蒸餾得來的
distinguish	区别，辨别	區別，辨別
distribute	分发，分配，散布，分布	分發，分配，散布，分布
distribution	分配，分发，分布	分配，分發，分布
disturb	弄乱，打乱，打扰，扰乱	弄亂，打亂，打擾，擾亂
diverge	道路分叉，意见分歧，脱离	道路分叉，意見分歧，脫離
divergence	分歧	分歧

English	Simplified Chinese	Complex Chinese
divergent	分歧的	分歧的
divide	分，划分，分开，隔开	分，劃分，分開，隔開
division	分开，分割，部门	分開，分割，部門
dizygote	二精合子	二精合子
DNA	脱氧核糖核酸	脫氧核糖核酸
DNase	脱氧核糖核酸酶	脫氧核糖核酸酶
dominant	有统治权的，支配的，显性的	有統治權的，支配的，顯性的
donor	捐赠者，供体	捐贈者，供體
door	门，通道，家，户	門，通道，家，戶
dorsal	背的，脊的	背的，脊的
doubling time	双倍时间	雙倍時間
drench	湿透	濕透
drink	饮料，酒，喝，喝酒，饮水	飲料，酒，喝，喝酒，飲水
dry (n, adj)	干的，干燥的，口渴的，干燥	乾的，乾燥的，口渴的，乾燥
dry ice	干冰	乾冰
dryness	干，干燥	乾，乾燥
duo	二重	二重
duplicate (n, adj)	重复的，复制的，副本，复写	重複的，複製的，副本，複寫
dust (n, v)	灰尘，尘土，尘埃，去掉灰尘	灰塵，塵土，塵埃，去掉灰塵
dysfunction	机能不良，功能紊乱	機能不良，功能紊亂
early	早的，早先的，早期的，初期	早的，早先的，早期的，初期
earth	地球，陆地，泥土，洞穴	地球，陸地，泥土，洞穴
ecology	生态学，均衡系统	生態學，均衡系統
ecosystem	生态系统	生態系統

English	Simplified Chinese	Complex Chinese
ectopic	异位的，异常的	異位的，異常的
editor	编辑，编辑器，编者	編輯，編輯器，編者
efficacy	功效，效验	功效，效驗
efficiency	功率，功效	功率，功效
egg	蛋，鸡蛋，卵，怂恿	蛋，雞蛋，卵，慫恿
eicosanoid	类花生酸	類花生酸
eighth	第八	第八
eightieth	第八十，八十分之一	第八十，八十分之一
electrical	电的， 有关电的	電的， 有關電的
electrode	电极	電極
electron microscope	电子显微镜	電子顯微鏡
electrophoresis	电泳，现象	電泳，現象
elevated	提高的，严肃的，欢欣的	提高的，嚴肅的，歡欣的
elevator	电梯，升降机，升降舵	電梯，升降機，升降舵
elongation	延长	延長
elution time	洗提时间	洗提時間
emaciate	使憔悴，消瘦	使憔悴，消瘦
e-mail	电子邮件	電子郵件
e-mail address	电子邮件地址	電子郵件地址
embryo	胚胎，胎儿，胚芽，胚胎的	胚胎，胎兒，胚芽，胚胎的
emerge	显现，浮现，暴露，形成	顯現，浮現，暴露，形成
emergency	紧急情况，突然事件	緊急情況，突然事件
emission	散发，发射，喷射	散發，發射，噴射
endocrine	内分泌	內分泌
endocytosis	内吞作用，吞噬作用	內吞作用，吞噬作用
endonuclease	核酸内切酶	核酸內切酶
endoplasmic	血浆内的	血漿內的

English	Simplified Chinese	Complex Chinese
endotoxin	内毒素	內毒素
English	英语，英文的，英国人	英語，英文的，英國人
enhancer	加强子，增强子，加强者	加強子，增強子，加強者
envelope	信封，封套，封袋	信封，封套，封袋
enzyme	酶	酶
enzyme linked immunoassay (ELIZA)	酶联免疫测定	酶聯免疫測定
epigenetic	外成的，后生的，表生的	外成的，後生的，表生的
episome	附加体，游离体，游离基因	附加體，游離體，游離基因
epistasis	上位性，异位显性	上位性，異位顯性
epithelium	上皮，上皮细胞	上皮，上皮細胞
equal (v, adj)	相等的，对手，匹敌，等于	相等的，對手，匹敵，等于
equivalent	相等的，同意义的，相等物	相等的，同意義的，相等物
erect (adj)	直立的，树立，建立	直立的，樹立，建立
ethanol	乙醇，酒精	乙醇，酒精
ether	乙醚	乙醚
euchromatin	常染色质	本染色質
evolution	进展，演变，进化	進展，演變，進化
exam	考试，测验	考試，測驗
examine	检查， 细查	檢查，細查
examination	考试，检查，细看	考試，檢查，細看
excision	切除，删除，切除术	切除，刪除，切除術
excitation	激发，兴奋，激励，鼓舞	激發，興奮，激勵，鼓舞
excite	刺激，使兴奋，使激动	刺激，使興奮，使激動
exhalation	呼出，呼气，蒸发，发散物	呼出，呼氣，蒸發，發散物
exhale	呼气，发出，发散，使蒸发	呼氣，發出，發散，使蒸發
exhaust (v)	用尽，耗尽，精疲力尽，排气	用盡，耗盡，精疲力盡，排氣

English	Simplified Chinese	Complex Chinese
existing	现有的，存在的	現有的，存在的
exocrine	外分泌的	外分泌的
exocytosis	胞外分泌	胞外分泌
exon	编码顺序，外显子	編碼順序，外顯子
exonuclease	核酸外切酶	核酸外切酶
expand	使膨胀，详述，扩张，发展	使膨脹，詳述，擴張，發展
explode	使爆炸，爆发，推翻，激发	使爆炸，爆發，推翻，激發
explosion	爆发，发出，爆炸	爆發，發出，爆炸
exponent	解释者，说明者，典型	解釋者，說明者，典型
exponential	指数，倡导者，幂数的	指數，倡導者，冪數的
expose	使暴露，受到，使曝光，揭露	使暴露，受到，使曝光，揭露
expression	表达，表情，表达式，符号	表達，表情，表達式，符號
extension	延长，扩充，范围，外延的	延長，擴充，範圍，外延的
exterminate	消除	消除
extinction coefficient	消光系数	消光系數
extra	额外的，特大的，特别的，非常	額外的，特大的，特別的，非常
extract (v)	榨，摘录，开方，析取，吸取	榨，摘錄，開方，析取，吸取
extrapolate	推断，外推	推斷，外推
eye protection	眼保护	眼保護
eye wash station	洗眼台(站)，洗眼地点	洗眼站，洗眼地點
face mask	面罩	面罩
factorial	因子的，阶乘的，阶乘	因子的，階乘的，階乘
fail	失败，不及格，忘记	失敗，不及格，忘記
false	错误的，假的，伪造的，欺诈的	錯誤的，假的，偽造的，欺詐的

English	Simplified Chinese	Complex Chinese
flame (n)	火焰，光辉，光芒，热情，激情	火焰，光輝，光芒，熱情，激情
flask	培养瓶	培養瓶
flight	飞行，班机，射击	飛行，班機，射擊
flocculent	羊毛状的，絮凝的，丛毛状的	羊毛狀的，絮凝的，叢毛狀的
floor mat	地板垫	地板墊
flower (n, v)	花，开花	花，開花
fluorescence	荧光，荧光的	螢光，螢光的
fog (n, v)	雾，烟雾，尘雾，使困惑	霧，煙霧，塵霧，使困惑
foil	箔，金属薄片，衬托	箔，金屬薄片，襯托
fold (v)	折，折叠，折叠起来	折，折疊，折疊起來
force (n, v)	力量，武力，军队，强制，夺取	力量，武力，軍隊，強制，奪取
forceps	镊子，钳子	鑷子，鉗子
fortieth	第四十，四十分之一	第四十，四十分之一
forward	早的，前进的，转寄，向前	早的，前進的，轉寄，向前
fourth	第四，四分之一	第四，四分之一
fractionate	使分馏	使分餾
frame	框架，结构，画面	框架，結構，畫面
frameshift	移码的,移码	移碼的, 移碼
free (n)	自由的，免费的，丰富的	自由的，免費的，豐富的
free (v)	释放，自由	釋放，自由
freeze	结冰，冷冻，冻结	結冰，冷凍，凍結
freezer	冷冻室，冷藏工人	冷凍室，冷藏工人
fume hood	排气罩	排氣罩
function	功能，官能，作用，函数，运行	功能，官能，作用，函數，運行
fungus	真菌, 菌类	真菌, 菌類

English	Simplified Chinese	Complex Chinese
g-forces	重力	重力
gain	得到，增进，获得，增加	得到，增進，獲得，增加
galleys	大型划船，军舰，校样	大型划船，軍艦,校樣
gallon	加仑	加侖
gamete	接合体，配偶子	接合體，配偶子
gamma radiation	γ辐射，γ放射	γ輻射，γ放射
ganglion	神经节	神經節
gap	缺口，裂口，缝隙，差距	缺口，裂口，縫隙，差距
gas	气体，煤气，毒气，汽油	氣體，煤氣，毒氣，汽油
gasket	垫圈，衬垫	墊圈，襯墊
gastrotoxic	胃毒的	胃毒的
gate	大门，入口	大門，入口
gauge (n)	标准尺，规格，量表，测量	標準尺，規格，量表，測量
GC	气相色谱	氣相色譜
gene	遗传因子，遗传基因	遺傳因子，遺傳基因
genesis	起源	起源
genome	基因组	基因組
genotoxic	遗传毒的	遺傳毒的
genotype	基因型	基因型
gland	腺体，腺	腺體，腺
global	球行的，全球的，全世界的	球行的，全球的，全世界的
gloves	手套，戴手套	手套，戴手套
glue	胶，胶水，粘合，粘贴	膠，膠水，粘合，粘貼
glutamic acid	谷氨酸	谷氨酸
glutamine	谷氨酰胺	谷氨酸鹽
glycan	多糖，聚糖，多聚糖	多糖，聚糖，多聚糖
glyceride	甘油酯，脂肪酸丙酯	甘油酯，脂肪酸丙酯

English	Simplified Chinese	Complex Chinese
glycerol	甘油，丙三醇	甘油，丙三醇
glycine	糖胶，氨基乙酸，甘氨酸	糖膠，氨基乙酸，甘氨酸
Golgi	高尔基	高爾基
gram	克	克
gram negative	革兰氏阴性	革蘭氏陰性
gram positive	革兰氏阳性	革蘭氏陽性
growth	生长，发育，栽培，种植	生長，發育，栽培，種植
growth factor	生长因子	生長指數
growth media	培养基	生長培養基
guanine	鸟嘌呤	鳥嘌呤
guarantee	保证，担保书，担保，抵押品	保證，擔保書，擔保，抵押品
guide (n)	指导，支配，带领，向导	指導，支配，帶領，向導
hair	头发，毛发，毛状物，绒毛	頭發，毛發，毛狀物，絨毛
half-time	半衰期	半衰期
Hanks' buffered saline	Hank 氏缓冲盐水	Hank氏緩沖鹽水
haploid	（染色体）单倍体，单一的	（染色體）單倍體，單一的
hapten	半抗原	半抗原
harmony	协调，融洽	協調，融洽
haze	薄雾，模糊，使糊涂	薄霧，模糊，使糊塗
head	头，顶点，朝向，出发，主要的	頭，頂點，朝向，出發，主要的
hearing	听力，倾听, 听证	聽力，傾聽, 聽證
heat (n)	热，热烈，高潮，压力	熱，熱烈，高潮，壓力
heat (v)	加热，使激动，发怒	加熱，使激動，發怒
heater	加热器，发热器	加熱器，發熱器
heating block	加热板	加熱板
heavy	重的，大量的	重的，大量的

English	Simplified Chinese	Complex Chinese
heterozygous	杂合子	雜合子
hexagon	六角形，六边形	六角形，六邊形
hide	隐藏，隐瞒	隱藏，隱瞞
high	高度，高原的，高度地	高度，高原的，高度地
histidine	组氨酸，组织氨基酸	組氨酸，組織氨基酸
histone	组蛋白	組蛋白
holder	支撑物，固定器	支撐物，固定器
home page	主页	主頁
homogenize	均质化，使均匀	均質化，使均勻
homogenous	同质的，纯系的	同質的，純系的
homology	同源性	同源性
homozygous	纯合子，同型结合的	純合子，同型結合的
horizontal	水平的，地平线的	水平的，地平線的
horseradish peroxidase	辣根过氧化物酶	辣根過氧化物酶
host (n)	宿主，主人	宿主，主人
hotspot	热点，热区	熱點，熱區
housekeeping	家务管理，卫生管理	家務管理，衛生管理
HPLC	高效液相色谱	高效液相色譜
human	人，人类	人，人類
humidity	湿度，潮湿	濕度，潮濕
hundredth	第一百的	第一百的
hybrid	混合物，杂交的	混合物，雜交的
hybridize (v)	杂交	雜交
hybridization (n)	杂交	雜交
hydrate (n)	氢氧化物	氫氧化物
hydrophilic	亲水的，吸水的	親水的，吸水的
hydrophobic	狂犬病的，疏水性的	狂犬病的，疏水性的

English	Simplified Chinese	Complex Chinese
hydrocortisone	氢化可的松	氫化考地松
Hz	赫兹	赫茲
ice	冰，结冰，冰冻	冰，結冰，冰凍
idea	想法，念头，意见，观点，概念	想法，念頭，意見，觀點，概念
idle	无用的，无价值的，浪费，虚度	無用的，無價值的，浪費，虛度
ignite	点火，点燃	點火，點燃
immediate	直接的，紧接的，立即的	直接的，緊接的，立即的
immobilize	固定不动	固定不動
immortalize	使不灭，使不朽，不老	使不滅，使不朽，不老
immune	免疫的	免疫的
immunity	免疫性	免疫性
immunoprecipitation	免疫沉淀	免疫沉澱
impart	给予，传授，告知，透露	給予，傳授，告知，透露
imperfect	有缺点的，未完成的，减弱的	有缺點的，未完成的，減弱的
implant	灌输, 植入	灌輸, 植入
imprint (v)	留下烙印	留下烙印
improve	改善，改进	改善，改進
in situ	原位	原位
incandescence	白热，炽热	白熱，熾熱
incline	倾斜，斜坡，倾向	傾斜，斜坡，傾向
incubation	孵育，熟虑	孵育，熟慮
incubator	培养箱	培養箱
index (n, v)	索引，指数，做索引	索引，指數，做索引
indirect	间接的，迂回的	間接的，迂回的
indoor	户内的，室内的	戶內的，室內的
infection	感染，传染，影响，传染病	感染，傳染，影響，傳染病

English	Simplified Chinese	Complex Chinese
inflame	燃烧	燃燒
inflammation	炎症，发炎	炎症，發炎
infrared	红外线的，红外的	紅外線的，紅外的
inhale	吸入，吸气	吸入，吸氣
inhibitor	抑制剂，抑制者	抑制劑，抑制者
initiate	开始，发动，传授	開始，發動，傳授
inject	注射，注入	注射，注入
ink	墨水	墨水
inkjet	喷墨	噴墨
innovate	革新，创造	革新，創造
innovative	创新的，革新的	創新的，革新的
insect	昆虫，虫的	昆虫，虫的
insert (n)	插入物	插入物
insert (v)	插入，嵌入，	插入，嵌入，
inspect	检查，视察	檢查，視察
inspiration	灵感 , 激励	靈感 , 激勵
instill	灌注	灌注
instruction	指示，说明书，指导	指示，說明書，指導
insult (v)	侮辱，凌辱	侮辱，凌辱
integral	完整的，整体的，(数学)积分的	完整的，整體的，(數學)積分的
integrate	使成整体，使一体化，结合	使成整體，使一體化，結合
intensity	强烈，剧烈，强度，亮度	強烈，劇烈，強度，亮度
interface	分界面，接触面，界面	分界面，接觸面，界面
interphase	细胞的间期，界间的，相间的	細胞的間期，界間的，相間的
interrogate	审问，询问，审问	審問，詢問，審問
introduction	介绍，导言，绪论	介紹，導言，緒論

English	Simplified Chinese	Complex Chinese
intron	基因内区	基因內區
inversion	颠倒	顛倒
invert	转化的，使颠倒，使转化	轉化的，使顛倒，使轉化
investigate	调查，研究	調查，研究
investigation	调查，研究	調查，研究
ion	离子	離子
ionophore	离子载体	離子載體
irradiate	照射	照射
irreproducible	不能繁殖的，不能复制的	不能繁殖的，不能複製的
isocratic	权利平等主义的	權利平等主義的
isoelectric point	等电离点	等電離點
isoleucine	异亮氨酸	異亮氨酸
isopropyl alcohol	异丙醇	異丙醇
journal	定期刊物，杂志	定期刊物，雜誌
karyotype	染色体组型	染色體組型
keyboard	键盘	鍵盤
kilogram	千克	千克
kinase	激酶	激酶
kit	试剂盒，工具包	試劑盒，工具包
knife	刀，餐刀	刀，餐刀
knob	把手，旋转	把手，旋轉
lab coat	实验室工作服	實驗室工作服
label (n, v)	标签，贴标签	標簽，貼標簽
laboratory	实验室	實驗室
lamp	灯，照亮	燈，照亮
laptop computer	手提电脑	手提電腦
laser	激光	激光

English	Simplified Chinese	Complex Chinese
late	迟的，新近的，晚，最近	遲的，新近的，晚，最近
lateral	侧部，支线，面的	側部，支線，側面的
law	法律，规则	法律，規則
leaf	叶，树叶，翻书页	葉，樹葉，翻書頁
leaflet	小页，传单，广告	小頁，傳單，廣告
leg	走，跑，小腿	走，跑，小腿
leucine	亮氨酸	亮氨酸
leukotriene	白三烯	白三烯
level (n)	水平，级别，标准	水平，級別，標準
lever	杆，杠杆，控制杆，抬起	杆，杠杆，控制杆，抬起
lid	盖子，眼睑	蓋子，眼瞼
lift	举起，电梯，升高，提高	舉起，電梯，升高，提高
ligand	配位子，配体	配位子，配體
light (adj)	轻的，发光的，明亮的	輕的，發光的，明亮的
light (n)	光，日光，发光体，灯	光，日光，發光體，燈
limited	有限的，狭窄的，缺乏创见的	有限的，狹窄的，缺乏創見的
line graph	线图	線圖
linear	线的，直线的，线性的	線的，直線的，線性的
lipid	脂质，油脂	脂質，油脂
lipooxygenase	脂(肪)氧合酶	脂(肪)氧合酶
liquid	液体，流体，流动的	液體，流體，流動的
liquid nitrogen	液氮	液氮
liter	公升	公升
lose	遗失，浪费，错过，受损失	遺失，浪費，錯過，受損失
loud	高声的，大声地	高聲的，大聲地
low	低的，低价，浅的	低的，低價，淺的
lower	较低的，下级的，减弱	較低的，下級的，減弱

English	Simplified Chinese	Complex Chinese
luminescence	发光	發光
luminol	发光氨	發光氨
lyophilize	使冻干	使凍乾
lysate	细胞裂解物	細胞裂解物
lyse	裂解	裂解
lysine	赖氨酸	賴氨酸
lysis	溶解，裂解	溶解，裂解
magnifier	放大镜，放大器	放大鏡，放大器
maintenance	维护，保持，生活费用	維護，保持，生活費用
malfunction	故障，机能失常，机能障碍	故障，機能失常，機能障礙
manual	手册，指南，手的，手工的	手冊，指南，手的，手工的
manuscript	手稿，原稿，草稿	手稿，原稿，草稿
margin	边缘，页边的空白	邊緣，頁邊的空白
marker	标记，标志，记号	標記，標志，記號
masculine (adj,n)	男性的，阳性的，男子	男性的，陽性的，男子
mass	大多数，质量，群众，聚集	大多數，群眾，质量，聚集
mass spectrometer	质谱仪	質譜儀
match (n)	火柴，比赛，匹配	火柴，比賽，匹配
match (v)	相配，相称，比赛，	相配，相稱，比賽
matter	事件，问题，物质，内容，原因	事件，問題，物質，內容，原因
maximum	最多量，最大限度，极大	最多量，最大限度，極大
mean	平均数	平均數，中間
measure	测量，测定，估量，调节	測量，測定，估量，調節
median	中位数，中值	中位數，中值
medium (n)	介质，培养液	介質，培養液
medium (adj)	中间的，中等的，半生熟的	中間的，中等的，半生熟的
meeting	会议，集会，会见	會議，集會，會見

English	Simplified Chinese	Complex Chinese
mega	百万，大	百萬，大
membrane	膜	膜
memory	记忆，储存	記憶，儲存
meniscus	新月形物，半月板，凸凹透镜	新月形物，半月板，凸凹透鏡
metabolic	代谢作用的，新陈代谢的	代謝作用的，新陳代謝的
metabolism	新陈代谢，变行	新陳代謝，變行
metabolite	代谢产物	代謝產物
metabolize	产生代谢变化	產生代謝變化
metal	金属	金屬
metallic	金属性的	金屬性的
meter	米，公尺，计，表	米，公尺，計，儀表
methanol	甲醇	甲醇
methionine	蛋氨酸，甲硫氨酸	蛋氨酸，甲硫氨酸
method	方法，办法	方法，辦法
methylene chloride	氯化亚甲基	氯化亞甲基
metric	米制的，公制的	米制的，公制的
micelle	胶束，胶囊，微胞，胶态离子	膠束，膠囊，微胞，膠態離離子
microbalance	微量天平	微量天平
microbial	微生物的，由细菌引起的	微生物的，由細菌引起的
microcentrifuge	微量离心机	微量離心機
microfilament	微丝，微纤丝	微絲，微纖絲
microfuge	微量离心	微量離心
microgram	微克	微克
micrometer	测微尺，千分尺，微米	測微尺，千分尺，微米
micromole	微摩尔，微克分子	微摩爾，微克分子
microscope	显微镜	顯微鏡

English	Simplified Chinese	Complex Chinese
migrate	移动，移植，迁居	移動，移植，遷居
millennium	太平盛世，一千年	太平盛世，一千年
milligram	毫克	毫克
millimeter	毫米	毫米
millimole	毫摩尔	毫摩爾
million	百万	百萬
millionth	第一百万，百万分之一	第一百萬，百萬分之一
mince	切碎	切碎
minimum	最小的，最低的，最小值	最小的，最低的，最小值
minus	负的，减的，减去，负数	負的，減的，減去，負數
mitochondria	线粒体	線粒體
mixer	混合器，搅拌器	混合器，攪拌器
mobility	活动性，灵活性，机动性	活動性，靈活性，機動性
mode	方式，模式，样式，时尚	方式，模式，樣式，時尚
modification	更改，修改	更改，修改
modify	修改	修改
moist	潮湿的，潮湿	潮濕的，潮濕
molarity	摩尔浓度	摩爾濃度
mole	摩尔	摩爾
molecular	分子的	分子的
molecular weight	分子量	分子量
monitor	监测，监控，监听器	監測，監控，監聽器
monozygote	纯合子	純合子
motile	运动的，显示活力的，运动型	運動的，顯示活力的，運動型
motor	发动机，电动机	發動機，電動機
multiply	繁殖，乘，增加	繁殖，乘，增加
muscle	肌肉	肌肉

English	Simplified Chinese	Complex Chinese
music	音乐，乐曲	音樂，樂曲
mutagen	诱变剂	誘變劑
mutagenic	诱变的	誘變的
mutation	突变	突變
mycoplasma	支原体，支原菌	支原體，支原菌
mycotoxin	霉菌毒菌(素)	霉菌毒菌(素)
myopic	近视的	近視的
myotoxic	肌毒的	肌毒的
n	数目	數目
nail	指甲，钉子，钉，钉牢	指甲，釘子，釘，釘牢
nanogram	毫微克，纤克	毫微克，纖克
nanometer	十亿分之一(10^{-9}) 公尺， 毫微(10^{-9})米	十億分之一(10^{-9})公尺，毫微(10^{-9})米
nanomole	毫微(10^{-9})摩尔，毫微摩	毫微(10^{-9})摩爾，毫微摩
narrow (adj)	狭窄，精密的，严密的，有限的	狹窄，精密的，嚴密的，有限的
neck	脖子，颈，颈状物	脖子，頸，頸狀物
needle	针	針
negative	否定，负数，阴性的，底片	否定，負數，陰性的，底片
nephrotoxic	肾脏毒的	腎臟毒的
nerve	神经，胆量，勇气	神經，膽量，勇氣
nervous	不安的，神经紧张的	不安的，神經緊張的
network	网络，网状物，广播网	網絡，網狀物，廣播網
neurotoxic	神经毒的	神經毒的
neutral	中立者，中性的	中立者，中性的
ninetieth	第九十，九十分之一	第九十，九十分之一
ninth	第九，九分之一	第九，九分之一
nitrate	硝酸盐	硝酸鹽

English	Simplified Chinese	Complex Chinese
nitrite	亚硝酸盐	亞硝酸鹽
nitrocellulose	硝基纤维素膜	硝基纖維素膜
noise	噪音，噪声	噪音，噪聲
non-parametric	非参数的	非參數的
non-specific	不精确的，不明确的，非特效的	不精確的，不明確的，非特效的
normality	当量浓度，正常状态	當量濃度, 正常狀態
Northern (RNA) blot	Northern RNA 印迹	Northern RNA 印跡
notebook	笔记本	筆記本
novel	小说，新奇的，新颖的，异常的	小說，新奇的，新穎的，異常的
nuclear	核子的，原子能的，中心的	核子的，原子能的，中心的
nucleotide	核苷酸	核苷酸
numerator	分子	分子
nut	坚果，螺母，螺帽，难解的问题	堅果，螺母，螺帽，難解的問題
obscure (adj)	暗的，朦胧的，模糊的	暗的，朦朧的，模糊的
observation	观察，观测，观察资料	觀察，觀測，觀察資料
observe	观察，观测，遵守，评述	觀察，觀測，遵守，評述
obstruct	阻塞，阻隔，障碍物	阻塞，阻隔，障礙物
obstruction	阻塞，妨碍，障碍物	阻塞，妨礙，障礙物
obtuse	钝的，迟钝的，愚蠢的	鈍的，遲鈍的，愚蠢的
occlude	使闭塞，使锢囚	使閉塞，使阻塞
occlusion	闭塞	閉塞
octagon	八边行	八邊形
odor	气味，名声	氣味，名聲
oncogene	致癌基因	致癌基因
oncogenesis	瘤形成	瘤形成
opening	开放的，开始，开幕的	開放的，開始，開幕的

English	Simplified Chinese	Complex Chinese
optic	眼的，视觉的，光学上的	眼的，視覺的，光學上的
orbital rocker	旋转混匀器	旋轉混勻器
organization	组织，机构，团体	組織，機構，團體
organize	组织，有机化，组织起来	組織，有機化，組織起來
original	最初的，原始的，原物，原作	最初的，原始的，原物，原作
ortho	垂直，直	垂直，直
osmolality	重量摩尔渗透压浓度	重量摩爾滲透壓濃度
osmolarity	摩尔渗透压浓度	摩爾滲透壓濃度
outdoor	室外的，户外的，野外的	室外的，戶外的，野外的
out-of-order	次序颠倒，不整齐，状态不好	次序顛倒，不整齊，狀態不好
oval	卵形的，椭圆的	卵形的，橢圓的
over	结束，越过，在……上，越过	結束，越過，在……上，越過
overhead	在头上的，高架的，在空中	在頭上的，高架的，在空中
oxidant	氧化剂	氧化劑
oxidize	氧化	氧化
packaging	包装	包裝
pad (n)	垫，衬垫	墊，襯墊
pain	痛苦，劳苦，努力	痛苦，勞苦，努力
palindrome	回文(顺读倒读都一样的词语)	回文(順讀倒讀都一樣的詞語)
paper	纸，报纸，文件，论文，用纸包装	紙，報紙，文件，論文，用紙包裝
para	保护，顺	保護，順
parabola	抛物线	拋物線
parallel	平行的，相同的，类似的，相应	平行的，相同的，類似的，相應
paralyze	使瘫痪，使麻痹	使癱瘓，使麻痹

English	Simplified Chinese	Complex Chinese
parametric	参数的，参变量的	參數的，參變量的
partial	部分的，局部的，偏袒的，偏爱的	部分的，局部的，偏袒的，偏愛的
particle	粒子，微粒	粒子，微粒
party	政党，社交活动	政黨，社交活動
pass	通过，途径，传递，流通	通過，途徑，傳遞，流通
passage	传代,通过，经过，通道，通路，节，段	傳代,通過，經過，通道，通路，節，段
patent (n, v)	专利，执照，特许的，显著的	專利，執照，特許的，顯著的
PCR	多聚酶链反应	多聚酶鏈反應
pellet (n)	小球，沉淀物	小球，沉澱物
pending	未决的	未決的
penetrance	渗透率	滲透率
percent	百分比，百分数	百分比，百分數
perception	感觉，知觉，理解	感覺，知覺，理解
perennial	四季不断的，终年的，长期的，永久的	四季不斷的，終年的，長期的，永久的
perfect (adj)	完美的，理想的，正确的，熟练的	完美的，理想的，正確的，熟練的
peripheral	末梢的，外围的	末梢的，外圍的
permanent marker	永久标记笔	永久標記筆
permeability	渗透性	滲透性
permeabilize	使渗透	滲透
permeable	有浸透性的，能透过的	有浸透性的，能透過的
permeate	弥漫，渗透，透过，充满	彌漫，滲透，透過，充滿
permission	许可，允许	許可，允許
permitted	许可了的，允许了的	許可了的，允許了的
peroxidase	过氧化物酶	過氧化物酶

English	Simplified Chinese	Complex Chinese
peroxide	过氧化物，过氧化氢	過氧化物，過氧化氫
perpendicular	垂直的，正交的，垂线	垂直的，正交的，垂線
perpetual	永久的	永久的
perspective	远景，前途，前瞻	遠景，前途，前瞻
pH buffer	pH 缓冲液	pH緩沖液
pH paper	pH 试纸	pH試紙
pharmacology	药理学	藥理學
pharmacy	药店,药房，药剂学，制药业	藥店,藥房，藥劑學，製藥業
phenotype	表现型，显型	表現型，顯型
phenylalanine	苯丙氨酸	苯基丙氨酸
phosphatase	磷酸酯酶	磷酸酯酶
phosphate	磷酸盐	磷酸鹽
phosphate buffered saline	磷酸盐缓冲液	磷酸鹽緩沖液
phospholipase	磷脂酶	磷脂酶
phospholipid	磷脂	磷脂
phosphorylation	磷酸化	磷酸化
phospho-specific	磷酸化专一性	磷酸化專用的
physiological	生理学的，生理学上的	生理學的，生理學上的
pie chart	圆图	圓圖
pint	品脱	品脫
pipette	移液管	移液管
pipette tip	移液管套头，枪头	移液管套頭，槍頭
pipettor	移液器	移液器
plant	植物，工厂，车间，设备	植物，工廠，車間，設備
plaque	噬菌斑	噬菌斑
plasma	血浆，乳浆	血漿，乳漿
plasmid	质粒，质体	質粒，質體

English	Simplified Chinese	Complex Chinese
plastic	塑料，整形	塑料，整形
plug (n, v)	插，塞，插头，插销	插，塞，插頭，插銷
plumb	铅锤，垂直的，垂直	鉛錘，垂直的，垂直
poison	毒药，毒害，中毒，放毒	毒藥，毒害，中毒，放毒
polycarbonate	聚碳酸酯	聚碳酸酯
polymer	聚合体	聚合體
polymerase	聚合酶	聚合酶
polymerase chain reaction	PCR，聚合酶链反应	PCR，聚合脢鏈反應
polymorphism	多样性，多态现象	多行性，多態現象
polypropylene	聚丙烯	聚丙烯
polystyrene	聚苯乙烯	聚苯乙烯
pore	毛孔，小孔，气孔	毛孔，小孔，氣孔
position (n)	位置，职位	位置，職位
position (v)	安置，决定……的位置	安置，決定……的位置
positive	肯定的，积极的，正的，阳的	肯定的，積極的，正的，陽的
post test	后续检验	後續檢驗
poster	海报，招贴	海報，招貼
potency	力量，效价，效力	力量，效價，效力
potential	潜在的，势的，位的，潜力，电压	潛在的，勢的，位的，潛力，電壓
potentiate (v)	加强	加強
potentiation (n)	加强	加強
pound	磅	磅
powder	粉末，尘土	粉末，塵土
power supply	电源	電源
precipitant	沉淀剂，仓促的	沉澱劑，倉促的
precipitate (n)	沉淀物	沉澱物

English	Simplified Chinese	Complex Chinese
pre-existing	早先存在的，本来就有的	早先存在的，本來就有的
pre-incubate	预先孵育	預先孵育
presbyopic	远视眼的，老花眼的	遠視眼的，老花眼的
presentation	介绍，陈述，赠送，表达	介紹，陳述，贈送，表達
pressure	压力，电压，压迫，强制，紧迫	壓力，電壓，壓迫，強制，緊迫
pre-treat	预先处理，预先刺激(细胞)	預先處理，預先刺激(細胞)
primary	第一	第一
printer	打印机，印刷工	打印機，印刷工
printout	打印输出	打印輸出
prion	蛋白感染素	蛋白感染素
procedure	程序，手续	程序，手續
process (n)	过程，作用，方法，步骤，处理	過程，作用，方法，步驟，處理
processor	处理器，处理者	處理器，處理者
program (n)	节目，程序，计划，纲要	節目，程序，計劃，綱要
projectile	弹射，发射的	彈射，發射的
projector	放映机，幻灯机	放映機，幻燈機
proline	脯氨酸	脯氨酸
proof	证据，试验，检验，校对	證據，試驗，檢驗，校對
proper	适当的，正确的，固有的	適當的，正確的，固有的
property	财产，所有物，性质，特性	財產，所有物，性質，特性
prostaglandin	前列腺素	前列腺素
protease	蛋白酶	蛋白酶
protection	保护	保護
protein	蛋白质，蛋白的	蛋白質，蛋白的
protein A	A 蛋白	A蛋白質
protein G	G 蛋白	G蛋白質

English	Simplified Chinese	Complex Chinese
prove	证明，证实，检验，考验	證明，證實，檢驗，考驗
provide	供应，供给，准备	供應，供給，準備
proximal	最接近的, 近端的	最接近的, 近端的
publication	出版物，出版，发行，公布，发表	出版物，出版，發行，公布，發表
pump (n, v)	泵，抽水机，抽水，抽吸	泵，抽水機，抽水，抽吸
put	放，摆，安置，投掷	放，擺，安置，投擲
PVDF	聚偏二氟乙烯	聚偏二氟乙烯
quadrillion	千的五次方	千的五次方
qualitative	性质上的，定性的	性質上的，定性的
quantitative	数量的，定量的	數量的，定量的
quantity	量，数量	量，數量
quart	夸脱	夸脫
quiet	静止的，宁静的，安静，平静	靜止的，寧靜的，安靜，平靜
rack	架，放在架上	架，放在架上
radiation badge	辐射剂量牌	輻射劑量牌
radioimmunoassay	放射免疫测定	放射免疫測定
radius	半径，范围，界限	半徑，範圍，界限
raise	上升，升高，提出，唤起	上升，升高，提出，喚起
RAM	随机存取存储器，随机存取内存	隨機存取存儲器，隨機存取內存
random	随机，任意，胡乱的	隨機，任意，胡亂的
range (v, n)	范围，排列，归类于，平行	範圍，排列，歸類于，平行
rank (n, v)	等级，排列，归类	等級，排列，歸類
rate limiting	限速性的	限速性的
ratio	比，比率	比，比率
razor blade	剃刀	剃刀

English	Simplified Chinese	Complex Chinese
readout	读出器，读数	讀出器，讀出
reaction	反应	反應
real-time	实时	實時
rearrange	再排列，重新安排	再排列，重新安排
receive	收到，接到，接收，接待	收到，接到，接收，接待
receptor	接受器，受体，感受器	接受器，受體，感受器
recessive	退行的，隐性的，隐性状的	退行的，隱性的，隱性狀的
recline	放置	放置
reconcile	使和解，使和谐，使顺从	使和解，使和諧，使順從
recover	恢复，痊愈，复原	恢復，痊癒，復原
rectangle	长方形，矩形	長方形，矩形
rectangular	矩形的，成直角的	矩形的，成直角的
reducing agent	还原剂	還原劑
reference (v, n)	提及，参考，参考书目，证明书	提及，參考，參考書目，證明書
refrigerate	使冷却，使清凉，冷藏	使冷卻，使清涼，冷藏
refuse (v)	拒绝，谢绝	拒絕，謝絕
regain	收回，恢复，重新夺得	收回，恢復，重新奪得
regulation	规则，规章，调节，校准	規則，規章，調節，校准
reject (v, n)	被拒之人，落选者，拒绝，抵制	被拒之人，落選者，拒絕，抵制
release (v)	释放，放弃，免除，发布	釋放，放棄，免除，發布
removal	移动，免职，切除	移動，免職，切除
renature	复性，使变性蛋白恢复	復性，使變性蛋白恢復
repair (v)	修理，修补	修理，修補
repeat (n, v)	重复，复制，使再现，反复	重複，複製，使再現，反復
replace	取代，替代，代替	取代，替代，代替
replicate	复制	複製

English	Simplified Chinese	Complex Chinese
replication	复制	複製
replicon	复制子（DNA 的组分）	複製子（DNA的組分）
reply (v)	答复，报复，答辩，回击，回答	答復，報復，答辯，回擊，回答
reprint (n)	再版，复印，单行本	再版，複印, 單行本
reproduce	繁殖，再生，复制	繁殖，再生，複製
reproducible	能繁殖的，可再生的，可复写的	能繁殖的，可再生的，可複寫的
rescue (v)	救援，营救	救援，營救
reset	重新安排，回置	重新安排, 回置
resistance	反抗，抵抗，阻力，电阻，阻抗	反抗，抵抗，阻力，電阻，阻抗
resistant	抵抗的，有抵抗力的	抵抗的，有抵抗力的
resolution	坚定，决心，决定，决议	堅定，決心，決定，決議
resolve (n, v)	决心，解决，溶解	決心，解決，溶解
respective	分别的，各自的	分別的，各自的
respire	呼吸	呼吸
respond	回答，响应，作出反应，有反应	回答，響應，作出反應，有反應
response	回答，响应，反应	回答，響應，反應
restrain	抵制，制止	抵制，制止
restraint	束缚	束縛
restriction digest	限制性消化	限制性消化
resuscitate	复苏，复兴	復蘇，復興
retainer	固定器	固定器
retention time	保留时间	保留時間
reticulum	网状组织	網狀組織
retrovirus	逆转录酶病毒	逆轉錄酶病毒
reverse (v, n)	相反，背面，反面，颠倒，	相反，背面，反面，顛倒，

English	Simplified Chinese	Complex Chinese
	倒转	倒轉
review (v, n)	回顾，复习，评论	回顧，復習，評論
review article	复习文献，综述	復習文獻, 綜述
rhythm	节奏，韵律	節奏，韻律
risk (n)	危险，风险	危險，風險
RNA	核糖核酸	核糖核酸
RNAse	核糖核酸酶	核糖核酸酶
rocker	混摇器	混搖器
roof	屋顶，顶	屋頂，頂
root (v, n)	根，根部，生根，坚定不移	根，根部，生根，堅定不移
rotor	转子	轉子
round off	完成，使完美	完成，使完美
RPM	每分钟转数	每分鐘轉數
rubber policeman	橡皮细胞刮脱器	橡皮細胞刮脫器
ruler	统治者，管理者，直尺，划线板	統治者，管理者，直尺，劃線板
safe (adj)	安全的，可靠的，有把握的	安全的，可靠的，有把握的
safety glasses	安全眼镜	安全眼鏡
sample (n, v)	样品，标本，例子，采样，尝试	樣品，標本，例子，採樣，嘗試
satellite	人造卫星	人造衛星
save	解救，保存，保留，节省	解救，保存，保留，節省
scalp	头皮	頭皮
scalpel	解剖刀	解剖刀
scanner	扫描仪	掃描儀
scanning	扫描	掃描
scientific notation	科学符号	科學符號
scissors	剪刀	剪刀

English	Simplified Chinese	Complex Chinese
score (v)	得分，记分，评价	得分，記分，評價
scrape	刮，擦，刮掉	刮，擦，刮掉
scraper	刮刀，刮器	刮刀，刮器
screw	螺丝钉，螺旋，调节	螺絲釘，螺旋，調節
screw cap	螺丝帽	螺絲帽
screwdriver	螺丝刀，螺丝起子	螺絲刀，螺絲起子
SDS PAGE	十二烷基磺酸钠聚丙烯酰胺凝胶电泳	十二烷基磺酸鈉聚丙烯酰胺凝膠電泳
second	秒，片刻，第二	秒，片刻，第二
secondary	第二	第二
section (n)	部分，断片，部件，节，项，页，地域	部分，斷片，部件，節，項，頁，地域
section (v)	切片	切片
secure (adj, v)	安全的，可靠的，放心的，保护	安全的，可靠的，放心的，保護
security	安全	安全
seed (n, v)	种子，萌芽，播种，结实	種子，萌芽，播種，結實
seminar	研究会，讨论会	研究會，討論會
send	送，寄，发送，派遣，打发	送，寄，發送，派遣，打發
sensitive	敏感的，灵敏的，感光的	敏感的，靈敏的，感光的
separate (adj, v)	分开的，分离的，分开，分别	分開的，分離的，分開，分別
sequence (n)	序列	序列
sequence (v)	测序	測序
serine	丝氨酸	絲氨酸
serum	血清，免疫血清	血清，免疫血清
server	服务器，服务员	服務器, 服務員
service (n)	服务，保养，维修	服務，保養，維修
seventh	第七	第七

English	Simplified Chinese	Complex Chinese
seventieth	第七十，七十分之一	第七十，七十分之一
shaft	轴，杆状物	軸，杆狀物
shake (v)	摇动，摇，颤抖，震动	搖動，搖，顫抖，震動
shaker	摇动器，混合器	搖動器，混合器
sham	佯装，伪品，虚假的	佯裝，偽品，虛假的
shelf	架子，搁板	架子，擱板
shield (n, v)	防护物，护罩，盾，盾状物，保护	防護物，護罩，盾，盾狀物，保護
shielded	被保护的	被保護的
shift (v)	移动，移位，转换，转移，改变	移動，移位，轉換，轉移，改變
shoe cover	鞋罩	鞋罩
show (v)	表示，展览，外观，显示	表示，展覽，外觀，顯示
shower (n)	淋浴，阵雨	淋浴，陣雨
shrink	收缩，缩短	收縮，縮短
sigmoidal	S 形曲线，S 形的	S形曲線，S形的
signal (n, adj)	信号，信号的	信號，信號的
signal (v)	发信号，用信号通知	發信號，用信號通知
significance	意义，重要性	意義，重要性
similar	相似的，类似的	相似的，類似的
simultaneous	同时的，同时发生的	同時的，同時發生的
sixth	第六，六分之一	第六，六分之一
sixtieth	第六十，六十分之一	第六十，六十分之一
sleeve	袖子	袖子
slice (n)	薄片，切片，部分	薄片，切片，部分
slice (v)	切	切
slick	光滑的，灵活的，打扮	光滑的，靈活的，打扮
slide (n)	幻灯片	幻燈片

English	Simplified Chinese	Complex Chinese
slide (v)	使滑动，使滑行	使滑動，使滑行
slippery	滑的，光滑的	滑的，光滑的
slit	切开，撕裂，裂缝	切開，撕裂，裂縫
slot (n)	缝，狭槽，细长的孔，足迹	縫，狹槽，細長的孔，足跡
slot (v)	开槽于，跟踪	開槽于，跟蹤
smell (n)	气味，嗅觉，臭味	氣味，嗅覺，臭味
snap cap	按式盖子	按式蓋子
software hardware	软件硬件	軟件硬件
soggy	浸水的，沉闷的	浸水的，沉悶的
solvent	溶解的，有溶解力的，溶媒，溶剂	溶解的，有溶解力的，溶媒，溶劑
sonicator	超声波仪	超聲波儀
sort (n, v)	种类，类别，分类，拣选	種類，類別，分類，揀選
sound (n, adj)	声音，可靠的，合理的，听起来	聲音，可靠的，合理的，聽起來
Southern (DNA) blot	Southern DNA 印迹	Southern DNA 印跡
spark	火花，闪烁，鼓舞，发动	火花，閃爍，鼓舞，發動
spatula	压舌板，药匙，刮铲	壓舌板，藥匙，刮鏟
speaker	演讲人，扬声器，喇叭	演講人，揚聲器，喇叭
specific	明确的，特别的	明確的，特別的
specimen	标本，样品，样本，待试验物	標本，樣品，樣本，待試驗物
spectrophotometer	分光光度计	分光光度計
spectroscopy	光谱学，波谱学	光譜學，波譜學
spend	花费，消耗，用尽，度过，消磨	花費，消耗，用盡，度過，消磨
sperm	精液，精子，鲸油	精液，精子，鯨油
splice (v)	接合，剪切	接合, 剪切

English	Simplified Chinese	Complex Chinese
spore	孢子，长孢子	孢子，長孢子
spot (n)	污点，地点，场所，现场	污點，地點，場所，現場
spot (v)	沾污，弄脏，侦察发现	沾污，弄臟，偵察發現
square	正方形，广场，平方的，四方的	正方形，廣場，平方的，四方的
square root	平方根	平方根
stage (n)	发展的进程，阶段或时期	發展的進程，階段或時期
stain (n, v)	污点，染污	污點，染污
stamen	雄蕊，雄性花蕊	雄蕊，雄性花蕊
standard	标准，规格，标准的，第一流的	標準，規格，標準的，第一流的
standard deviation	标准差	標準差
standard error	标准误	標準誤
staple (n)	钉书钉，原材料，分类	釘書釘，原材料，分類
start (v)	动身，开始，出发，着手	動身，開始，出發，著手
starve	(细胞)乏生长因子培养	(細胞)乏生長因子培養
statistical significance	统计学显著性意义	統計學顯著性意義
statistics	统计学，统计表	統計學，統計表
steam (n)	蒸汽，水汽，蒸发，蒸	蒸汽，水汽，蒸發，蒸
sterile	消毒的，不育的	消毒的, 不育的
sterile hood	消毒罩	消毒罩
sterilize	杀菌，消毒	殺菌，消毒
stir	搅动，搅和	攪動，攪和
stool	粪便，厕所	糞便，廁所
stop	停止，障碍，阻滞，终止，难倒	停止，障礙，阻滯，終止，難倒
strand (n)	线的一股，链	線的一股，鏈
strand (v)	搁浅，使落后，弄断	擱淺，使落後，弄斷

English	Simplified Chinese	Complex Chinese
stratosphere	同温层，最上层，最高阶段	同溫層，最上層，最高階段
stress (n, v)	重压，压力，强调，重读	重壓，壓力，強調，重讀
sublimate	升华，纯化	升華，純化
subtract	减去，减	減去，減
supernatant	上清液	上清液
superoxide	过氧化物，超氧化物	過氧化物，超氧化物
supine	仰卧的，掌心向上的，向后靠的	仰臥的，掌心向上的，向後靠的
supplement	补充，附录，增刊	補充，附錄，增刊
supplementation	增补，补充，追加	增補，補充，追加
support (v)	支撑，支持，支援，拥护，维持	支撐，支持，支援，擁護，維持
susceptibility	易感性	易感性
swell	膨胀，变大	膨脹，變大
swinging bucket	摇动水桶	搖動水桶
symposium	座谈会，讨论会	座談會，討論會
synchronize	同步	同步
synergy	协作，协同	協作，協同
synthase	合酶	合酶
synthesis	综合，合成	綜合，合成
synthetase	合成酶	合成酶
synthetic	合成的，人造的，综合的	合成的，人造的，綜合的
syringe	注射器，注射，冲洗，灌洗	注射器，注射，沖洗，灌洗
T150 flask	T150 培养瓶	T150 培養瓶
T25 flask	T25 培养瓶	T25培養瓶
T75 flask	T75 培养瓶	T75培養瓶
table	桌子，工作台，表格，指表	桌子，工作台，表格，指表
tachometer	转速计，流速计	轉速計，流速計

English	Simplified Chinese	Complex Chinese
tachymeter	准距仪	準距儀
tandem	一前一后	一前一後
tape (n, v)	带子，录音带，磁带，录音	帶子，錄音帶，磁帶，錄音
task	任务，作业，分派任务	任務，作業，分派任務
taste (v)	品尝，辨味，体验，感到	品嘗，辨味，體驗，感到
technique	技术，技巧，方法，表演法	技術，技巧，方法，表演法
teflon	聚四氟乙烯	聚四氟乙烯
temperature	温度	溫度
tera	万亿，兆兆	萬億，兆兆
teratogen	致畸剂，畸胎剂	致畸劑，畸胎劑
teratogenic	致畸的	致畸的
terminate	停止，结束，终止	停止，結束，終止
test (v)	测验，试验，检验	測驗，試驗，檢驗
test tube	试管	試管
tether (n)	范围，束缚	範圍, 束縛
tetrahedron	四面体	四面體
thin layer chromatography	薄层色谱法	薄層色譜法
third	第三，三分之一	第三，三分之一
thirtieth	第三十，三十分之一的	第三十，三十分之一的
thousandth	第一千的，千分之一的	第一千的，千分之一的
threonine	苏氨酸	蘇氨酸
thromboxane	凝血烷	凝血烷
through	经由，穿过，直达的	經由，穿過，直達的
thymidine	胸腺嘧啶脱氧核苷	胸腺嘧啶脫氧核苷
tie (v)	带子，线，打结	帶子，線，打結
tile (n, v)	瓦片，瓷砖	瓦片，瓷磚
time (n)	时间，时候，时机，次数	時間，時候，時機，次數

English	Simplified Chinese	Complex Chinese
time (v)	安排时间，记录时间，计时	安排時間，記錄時間，計時
timer	定时器，计时员，记时员	定時器，計時員，記時員
tissue	组织	組織
ton	吨	噸
tone (n)	音调，音质	音調，音質
touch (v)	接触，碰，涉及，提到	接觸，碰，涉及，提到
toxic	有毒的	有毒的
toxin	毒素	毒素
trademark (n)	商标	商標
trans	反式，会报，论文集，学术会议录	反式,，論文集，學術會議錄
transcriptase	转录酶	轉錄酶
transcription	转录，抄写	轉錄，抄寫
transduction	转换，换能	轉換，換能
transfection	转染	轉染
transfer (v)	转移，传递	轉移，傳遞
transferase	转移酶	轉移酶
transform	转换，改变	轉換，改變
transition	转变，转换，跃迁，变调	轉變，轉換，躍遷，變調
translation	翻译，平移	翻譯，平移
translocate	移位，改变位置	移位，改變位置
transmission	播发，发射，传送，传动，传输	播發，發射，傳送，傳動，傳輸
transversion	转换，颠换	轉換，顛換
tray	盘，碟，盘子	盤，碟，盤子
triangle	三角形	三角形
triangular	三角形的	三角形的
trim (adj)	整齐的，整洁的	整齊的，整潔的

English	Simplified Chinese	Complex Chinese
trio	三重唱	三重唱
triplet	三倍的	三倍的
triplicate	一式三份	一式三份
troposphere	对流层	對流層
troubleshoot	查找及排除故障，故障检定	查找及排除故障，故障檢定
true	真实的，正确的	真實的，正確的
try (v, n)	尝试，试图，努力，试验	嘗試，試圖，努力，試驗
trypsin	胰蛋白酶	胰蛋白酶
tryptophan	色氨酸	色氨酸
t-test	t-检验	t-檢驗
tubing (n)	装管，管道系统，管行材料	裝管，管道系統，管行材料
tubing (adj)	管状的，管制的，制管的	管狀的，管制的，制管的
tubule	小管，细管	小管，細管
turnover rate	周转率	週轉率
tweezers	镊子	鑷子
twentieth	第二十的	第二十的
twin	成对，孪生	成對，孿生
typical	典型的，象征性的	典型的，象徵性的
tyrosine	酪氨酸	酪氨酸
ultra	超，过度，限外	超，過度，限外
ultra freezer	超低温冰箱	超低溫冰箱
ultrabalance	微量天平	微量天平
unavailable	难以获得的	難以獲得的
under	在……下，低于	在……下，低于
underneath	在下面，在……的下面	在下面，在……的下面
unit	个体，计量单位	個體，計量單位
unity	团结，联合，统一	團結，聯合，統一

English	Simplified Chinese	Complex Chinese
unscheduled	未计划的	未計劃的
uridine	尿嘧啶核苷	尿嘧啶核苷
UV	紫外线	紫外線
vacuum (n, adj)	真空，真空的	真空，真空的
vacuum trap	负压吸引瓶	負壓吸引瓶
valine	缬氨酸	纈氨酸
vapor	蒸发	蒸發
variable	变数，可变物，变量，可变的	變數，可變物，變量，可變的
variance	变化，变异	變化，變異
variant	变量	變量
vector	载体，标量	載體，標量
vehicle	交通工具，媒介	交通工具，媒介
ventral	腹部的	腹部的
verify	校验，校验，查证，核实	校驗，校驗，查證，核實
vertical	垂直的，直立的，垂直线，竖向	垂直的，直立的，垂直線，豎向
vesicle	囊，泡，小泡	囊，泡，小泡
viability	生存能力，发育能力	生存能力，發育能力
viable	能养活的，能生育的，可行的	能養活的，能生育的，可行的
villus	绒毛	絨毛
viroid	植物致命的类病毒	植物致命的類病毒
virulent	剧毒的，致命的，恶性的	劇毒的，致命的，惡性的
virus	病毒，毒害，恶毒	病毒，毒害，惡毒
visible	可见的	可見的
vision	视力，视觉，景象	視力，視覺，景象
Volt	伏特	伏特
voltage	电压，伏特数	電壓，伏特數

English	Simplified Chinese	Complex Chinese
volume	卷，册，体积，量，大量，音量	卷，冊，體積，量，大量，音量
vortex mixer	旋涡混匀	旋渦混勻
warm (adj, n)	温暖的，兴奋的，保暖物	溫暖的，興奮的，保暖物
warranty	担保，保证，理由	擔保，保證，理由
wart	疣	疣
wash (v)	洗涤，冲洗	洗滌，沖洗
washer	洗衣机，洗碗机	洗衣機，洗碗機
waste (n, v)	废物，损耗，多余的	廢物，損耗，多餘的
water (n, v)	水，喷淋，流泪，加水，浇水	水，噴淋，流淚，加水，澆水
wavelength	波长	波長
web page	网页	網頁
Western (protein) blot	蛋白印迹	蛋白印跡
wet (adj)	湿的，潮湿的	濕的，潮濕的
wetness	湿润，下雨	濕潤，下雨
wheel (n)	轮，轮子	輪，輪子
whole	所有的，全部	所有的，全部
wide	宽的，广阔的，广大地	寬的，廣闊的，廣大地
wind (n)	风	風
wind (v)	绕，缠绕，旋紧	繞，纏繞，旋緊
window	窗	窗
wire (n)	金属丝，电信	金屬絲，電信
withdraw	收回，撤销，撤退	收回，撤銷，撤退
withhold	停止，拒给，保留	停止，拒給，保留
wood	木头	木頭
World Wide Web	国际互联网	國際互聯網
worse	更坏的，更恶劣的	更壞的，更惡劣的

English	Simplified Chinese	Complex Chinese
worsen	恶化，损害	惡化，損害
worst	最坏的，最差的	最壞的，最差的
wrench (n, v)	扳钳，痛苦，扭伤，曲解，折磨	扳鉗，痛苦，扭傷，曲解，折磨
X-ray	X射线	X射線
yeast	酵母，发酵粉	酵母，發酵粉
zygote	受精卵，接合子	受精卵，接合子

Illustrations Index

A

384 well assay plate, 26
96 well assay plate clear, 26
96 well assay plate opaque black, 26
96 well assay plate opaque white, 26
Adjustable micropipetter, 25
Analytical balance, 27
Antistatic brush, 27
Autoclave, 58
Autosampler vial, 51

B

Benchtop ultracentrifuge, 31
Beta emitter radiation shield, 58
Beta storage container, 58
Biohazard container, 58
Blotting membranes, 46
Blotting paper, 46
Bottle brush, 55
Bunsen burner, 39

C

Calculator, 53
Caliper, 28
Cell scraper, 39
Chromatogram, 50
Chromatography columns, 50
Conical bottom culture tube, 31
Connector, 54
Cordless pipetter, 25
Cryogenic gloves, 57
Cryogenic vials, 31
Culture dishes, 39
Culture flask, 39
Culture hood, 39
Culture media, 39
Culture shaker, 39
Culture vessel, 39
Cuvette, 26

D

Dessiccator, 34
Dishwashing soap, 55
Disposable syringe, 53
Disposable weighing dishes, 27
Dust mask, 57

E

Ear muff, 57
Ear plugs, 57
Electrical receptacle, 53
Electrodes, 51
Eyewash station, 58

F

Fire extinguisher, 58
First aid kit, 58
Fixed angle rotor, 31
Flammable safety cabinet, 58
Flowmeter, 55
Forceps, 54
FPLC system, 50
Fraction collector, 51
Freezer rack, 33
Full size ultracentrifuge, 31
Fume hood, 57
Funnel, 53

G

Gas cylinder support, 55
Gas tank regulator, 55
Gas-tight syringe, 53
Gel combs, 51
Gel documentation system, 43
Gel dryer, 34
Glass stoppers, 53

H

Hand protector, 57
Hazard label, 58
Heat gun, 32
Heating bath, 32

Heating block, 32
Horizontal gel electrophoresis system, 50
Hot plate, 32

I

Ice bucket, 32
Ice machine, 55
Ice scoop, 55
Incubator, 39
In-line air filter, 34
Inverted microscope, 46

L

Lab sink, 55
Lab stool, 53
Labeling tape, 54
Laboratory coat, 57
Laboratory notebook, 43
Laser pointer, 53
Latex glove, 57
Liquid nitrogen storage tank, 32
Loop, 39
Lyophilizer, 34

M

Markers, 54
Microarray scanner, 46
Microcentrifuge tubes, 31
Microcentrifuge, 31
Microliter syringe, 53
Micropipetter tips, 25
Microplate reader, 26
Microscope slides, 46
Microwave oven, 32
Mini gel system, 50
Multichannel pipetter, 26

O

Outlet strip, 53
Oven, 32

P

Paraffin film, 54
Pasteur pipette, 25
PCR plate, 47
PCR thermocycler, 47
pH meter, 27
pH testing strips, 27
Pipette bulb, 25
Pipettes, 25
Plug cap tube, 31
Power supply, 50
Pre-cast gel ready, 51

R

Radiation monitor, 57
Razor blade, 54
Reference weight set, 27
Refrigerating circulator, 32
Refrigerator/freezer, 33
Repeating pipetter, 26
Respirator, 57
Rocking platform, 33
Round bottom culture tube, 31
Rubber gloves, 57
Rubber stopper, 53

S

Safety glasses, 57
Safety goggles, 57
Safety shower, 58
Scalpel, 54
Scintillation counter, 28
Scissors, 54
Sequencing gel electrophoresis system, 50
Shoe covers, 57
Sink sprayer, 58
Slot blot manifold, 46
Snap cap tube, 31
Sonicating bath, 33
Sonicating probe, 33
Spatula, 27
Spectrophotometer, 26
Spill containment kit, 58
Squirt bottles, 55
Stereo dissecting microscope, 46
Sterile filtration unit, 34
Stir bars, 33
Stir plate, 33
Strip chart recorder, 43
Swinging bucket rotor, 31
Syringe needle, 53

T

Thermometer, 27
Thermomixer, 32
Timer, 28
Tissue grinder, 33
Transfer apparatus, 50
Tray, 53
Tubing clamps, 54
Tubing connector, 54
Tubing stopcock, 54
Tubing, 55
Tumbling mixer, 33

U

Ultra freezer, 33
Upright microscope, 46
UV/visible transilluminator, 43

V

Vacuum filtration unit, 34
Vacuum grease, 54
Vacuum pump, 54
Vacuum trap, 39
Vertical electrophoresis system, 50
Voltmeter, 28
Vortex mixer, 33

About the Authors

James M. Samet holds a PhD in toxicology and an MPH in environmental sciences from the University of North Carolina at Chapel Hill. He is currently a research biologist with the National Health and Environmental Effects Research Laboratory, and an Adjunct Assistant Professor in the Curriculum of Toxicology and the Department of Environmental Sciences at the University of North Carolina. His research focuses on signaling mechanisms activated by environmental pollutants.

Weidong Wu graduated from Henan Medical University with an MD and from Beijing Medical University with a PhD in Toxicology. Since 1998, he has worked at the University of North Carolina Center for Environmental Medicine and Lung Biology. He is presently a Research Assistant Professor in the Department of Pharmacology. His research interests include the cellular and molecular mechanisms of lung injury of inhaled toxicants.

Yuh-Chin T. Huang graduated with an MD from the Medical College of National Taiwan University. He also received a Master of Health Science degree from the School of Public Health at the Johns Hopkins University. He is currently a medical officer with the National Health and Environmental Effects Research Laboratory of the US Environmental Protection Agency, and a faculty member in the Division of Pulmonary and Critical Care Medicine at Duke University Medical Center and the University of North Carolina Memorial Hospital.

Xinchao Wang is a graduate of Henan Medical University, where he received his MD and MPH. He worked at the University of North Carolina for the last three years as a post-doctoral fellow. He is presently an Associate Professor in the Department of Occupational Medicine and Toxicology of Zhengzhou University, People's Republic of China. His research focuses on the effects of environmental toxicants on lung injury.